W. Kropatsch
R. Klette
F. Solina (eds.)
in cooperation with
R. Albrecht

Theoretical Foundations
of Computer Vision

Computing Supplement 11

Springer-Verlag Wien GmbH

Prof. Dr. W. Kropatsch

Institute for Pattern Recognition
and Image Processing
Technical University Vienna
Austria

Prof. Dr. F. Solina

Faculty of Electrical Engineering
and Computer Science
University of Ljubljana
Slovenia

Prof. Dr. R. Klette

Computer Science Department
Technical University at Berlin
Germany

Prof. Dr. R. Albrecht

Institute for Informatics
University of Innsbruck
Austria

©1996 Springer-Verlag Wien
Originally published by Springer-Verlag/Wien in 1996

Typesetting: Asco Trade Typesetters Ltd., Hong Kong

Printed on acid-free and chlorine-free bleached paper

With 87 Figures

Library of Congress Cataloging-in-Publication Data

Theoretical foundations of computer vision / W. Kropatsch, R. Klette,
F. Solina, eds. in cooperation with R. Albrecht.
      p.   cm. – (Computing supplement, ISSN 0344-8029 ; 11)
   Includes bibliographical references.
   ISBN 978-3-211-82730-7   ISBN 978-3-7091-6586-7 (eBook)
   DOI 10.1007/978-3-7091-6586-7

   1. Computer vision.   2. Image processing – Digital techniques.
I. Kropatsch, W. (Walter)   II. Klette, Reinhard.   III. Solina,
Franc.   IV. Series: Computing (Springer-Verlag).   Supplementum ; 11.
TA1634.T43   1996
006.4'2 – dc20   95-45033

ISSN 0344-8029

ISBN 978-3-211-82730-7

Preface

Computer Vision is a rapidly growing field of research investigating in computational and algorithmic issues associated with image acquisition, processing, and understanding. It serves tasks like manipulation, recognition, mobility, and communication in diverse application areas such as manufacturing, robotics, medicine, security and virtual reality. The development of the field was driven very much by concrete applications resulting in many individual but very specific solutions working satisfactory but mostly in a very restricted environment requiring a deep insight in the programs to adapt them to even small changes in the environment. So far only few systems can handle unforeseen situations, the space and the conditions of application are seldom well defined such that the re-use of software often relies mainly on trial and error.

A sound theoretical foundation of computer vision should help to overcome the above-mentioned problems. In particular it should contribute (1) to structure the scientific field; (2) to formulate theories that relate properties of real world objects with the methods to extract the required information from images taken under different and often varying and uncertain sensing situations; and (3) to a more general understanding of vision processes.

This supplementary volume contains a selection of papers devoted to theoretical foundations of computer vision. They have been presented and intensively discussed at the seventh workshop on this topic in the castle of Dagstuhl in March 1994. The subjects of the revised contributions in this volume cover a broad range of fields: e.g. motion analysis, discrete geometry, computational aspects of vision processes, models, morphology, invariance, image compression, 3D reconstruction of shape. Still no claim for completeness is made, however, several issues have been identified that are of essential interest to the community:

- non-linear operators;
- the transition between continuous to discrete representations;
- a new calculus of non-orthogonal partially dependent systems (R. Bajcsy).

Finally it should be mentioned that the workshop and this volume brought together scientists from both Western and Eastern Europe as well as the United States that helped to better understand the different views in the vision community.

March 1995   *Walter G. Kropatsch, Reinhard Klette, Franc Solina*

Contents

Computing Suppl. 11, 1–20 (1996)

Attentive Visual Motion Processing: Computations in the Log-Polar Plane

K. Daniilidis, Kiel

Abstract

Attentive Visual Motion Processing: Computations in the Log-Polar Plane. Attentive vision is characterized by selective sensing in space and time as well as selective processing with respect to a specific task. Selection in space involves the splitting of the visual field in a high resolution area—the fovea—and a space-variant resolution area—the periphery. Both in neurobiology and in robot vision, models of the resolution decrease towards the image boundaries have been established. The most convincing model is the theory of log-polar mapping where very high data compression rates are achieved. In combination with the complexity reduction we believe that the log-polar mapping has further computational advantages which we elaborate in this study. Based on the optical flow we study the computation of 3D motion and structure globally and locally. We present a global method to compute the Focus of Expansion in the case of pure translation. By fixating on an object we show how to estimate ego motion in the presence of translation and rotation of the observer from the flow in the log-polar periphery. Then, we turn to local differential computations and we establish both approximate and exact expressions for the time to collision.

Key words: Space-varient sensing, motion estimation, fixation.

1. Introduction

It is well known that when a human observer has to accomplish a navigational task like car driving the retinal information of the entire field of view is used for the estimation of 3D-motion and the control of the vehicle. It is also well known and easy to show experimentally that there is an immense decrease of the resolution towards the periphery of the retina. However, the coarse resolution on the major part of the field of view does not hinder the observer to detect alarming events and to exploit the image periphery for the estimation of ego-motion. In the same scenario, if the car-driver approaches a crossing he tries to estimate the relative motion of the cross-moving cars in order to react appropriately. Then the driver directs his gaze towards the moving object and pursues it for a short time interval. Such a gaze change and holding are carried out with a navigational and not a recognition purpose: the driver needs the information how the cross-moving car is moving rather than how it looks like.

In this paper, we study the properties of the motion field using a space variant resolution scheme, and we propose algorithms for the recovery of 3D-motion which exploit the advantages of space variant sensing and fixation. Space variant sensing is one of the main features of an attentive vision system. Attention arises as a necessity in building robot systems that are able to react according to the visual stimuli in a dynamically changing environment. Even with the fastest image processing architectures it is not possible to process a uniformly sampled image with a wide field of view in a process time enabling immediate reaction. To overcome the reaction time constraints we have to focus on a region of interest. However, even if we select a subset of the retinal information we have still to process the image data outside the focus of attention to detect new events. We, thus, need an image data reduction scheme which conserves high resolution in the center of the image—the fovea—and a gradually decreasing resolution area—the periphery. Since most visual motion tasks necessitate a wide field of view we have to elaborate algorithms for computing the optical flow and recovering the 3D-motion and structure using space-variant resolution. In this paper, we use the complex logarithmic mapping (or log-polar transformation) to model the non-uniform periphery of an image, and we keep the cartesian model for the uniform fovea. However, we do not only elaborate motion algorithms regarding the logarithmic polar plane as a practical necessity. We show that the log-polar mapping has also computational advantages with respect to motion recovery tasks:

- We propose an algorithm for the computation of the translation direction in case of pure translation using only 4% of the original data.
- We show that fixation enables the computation of the translation direction in case of general motion. Pursuit and optokinetic eye movements have long been considered by the psychophysicists [6] as a means to overcome the limited field of view, to bound the binocular disparity, and to eliminate the motion blur. Here, we prove that eye movements are not only an implication of the inhomogeneous retina, but they support the recovery of 3D-motion.
- In analogy to the approach in [23] we prove that the polar motion field enables the local computation of bounds to the time to collision which are independent of rotation and the slant of the viewed surface.

We proceed with a survey of approaches to space variant sensing and fixation. The first experimental findings on the density of the photoreceptors in the human retina are due to Schulze [18]—for a comparison of further biological findings the reader is referred to [6, 25]. Space variant arrangements concern not only the density of the photoreceptors but the geometry of the receptive fields in the cortex, too. Such retino-cortical mappings have been explicitly formulated by Schwartz [19] for cortical areas of the monkey and by Mallot [13] for cortical areas of the cat. Schwartz [19] proposed the complex logarithmic mapping and showed its invariance properties. The amount of literature on biological work is very large and we will not attempt to give a further review.

Weiman and Chaikin [28] studied the properties of the complex logarithmic transformation as a conformal mapping and they proposed a logarithmic spiral grid as a digitization scheme for both image synthesis and analysis. Sandini et al. [14] studied the recognition of 2D-patterns in the log-polar plane and they proposed optimality criteria for the subsampling of the non-foveal part of the image. The description of the first CCD-sensor with a cartesian fovea and a log-polar periphery can be found in [24]. Wallace et al. [26] implemented the log-polar mapping on a DSP-architecture by averaging sets of CCD pixels. They integrated the log-polar mapping into a miniaturized pan-tilt camera enabling a visual communication using voice-bandwidth channels.

Jain et al. [10] were the first who applied the complex logarithmic mapping in image sequence analysis. Given the motion information from other sensors and assuming pure translation they transformed the images into the log-polar plane using the focus of expansion as the center of the transformation. The depths of the viewed point were easily obtained by inspecting only the shifts along one coordinate of the log-polar plane. Tistarelli and Sandini [22, 23] derived the motion equations for the log-polar plane and proposed a method for the computation of the time to collision. Our work on local computations (Section 4) is inspired by their approach. The combination of fixation and space variant sensing appears only in [22]. The following references concern fixation in the cartesian plane. Aloimonos et al. [1] and Bandopadhay and Ballard [2] showed how the structure from motion problem is simplified in case of active tracking. Fermüller and Aloimonos [8] proved that normal flow measurements are sufficient for bounding the locus of the focus of expansion and the time to collision. Raviv and Herman [17] studied the geometric properties of fixation and derived the locus of 3D-points whose projections yield a zero-optical flow in case of motion with three degrees of freedom (road following). Taalebinezhaad [21] proposed a method that simulates tracking by a pixel shifting process without actively controlling the mechanical degrees of freedom of the camera. The translation direction is then obtained by minimizing the depth deviation from the fixation point using only normal flow measurements. A similar method was proposed in [9] for the derotation of the cartesian motion field.

The organization of the paper is as follows. First, we formally describe the complex logarithmic mapping. We derive the equations of the motion-field in the log-polar plane in Section 3. The rest of the study is divided to global motion computations (subsections of §3) and local parallax computations (Section 4).

2. The Complex Logarithmic Mapping

We use (x, y) for the cartesian coordinates and (ρ, η) for the polar coordinates in the plane. By denoting with $z = x + jy = \rho e^{j\eta}$ a point in the complex plane the complex logarithmic (or log-polar) mapping is defined as

$$w = \ln(z) \tag{1}$$

for every $z \neq 0$ where $\mathrm{Re}(w) = \ln \rho$ and $\mathrm{Im}(w) = \eta + 2k\pi$. To exclude the periodicity of the imaginary part we constrain the range of $\mathrm{Im}(w)$ to $[0, 2\pi)$. The complex logarithmic mapping is a well-known conformal mapping preserving the angle of

K. Daniilidis

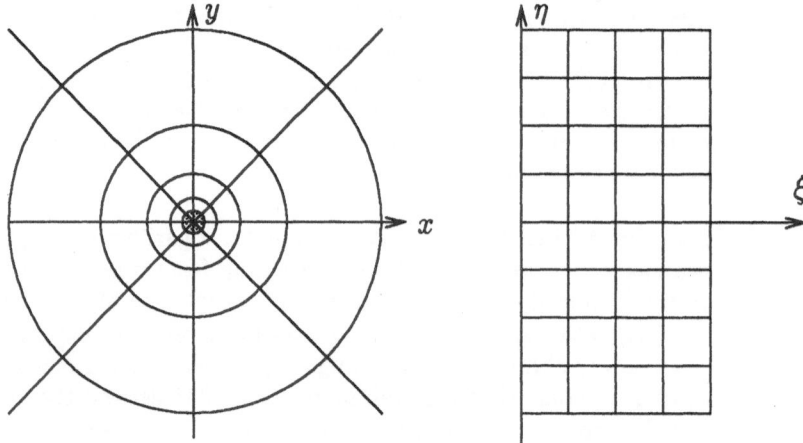

Figure 1. The complex logarithmic mapping maps radial lines and cocentric circles into lines parallel
to the coordinate axes. The log-polar variables ξ and η are defined in Eq. (2)

the intersection of two curves. It is illustrated in Fig. 1 where ξ denotes $\ln \rho$. It is
trivial to show that every scaling and rotation about the origin in the z-plane is
represented in the w-plane by a shift parallel to the real and imaginary axis,
respectively.

We apply the log-polar mapping on the non-foveal part of a retinal image. Therefore, we define as the domain of the mapping the ring-shaped area $\rho_0 < \rho < \rho_{max}$
where ρ_0 and ρ_{max} are the radius of the fovea and the half-size of the retinal image,
respectively. Furthermore, a hardware CCD-sensor with the log-polar property or
a software implementation of the mapping needs a discretization of the w-plane—
which we will call log-polar plane in contrast to the cartesian plane. By assuming
that N_r is the number of cells in the radial direction and N_a is the number of cells
in the angular direction the mapping from the polar coordinates (ρ, η) to the
log-polar coordinates (ξ, γ) reads as follows (see also [22])

$$\xi = \log_a \left(\frac{\rho}{\rho_0} \right)$$

$$\gamma = \frac{N_a}{2\pi} \eta \tag{2}$$

where the logarithmic basis a is obtained from the foveal radius ρ_0, the image
radius ρ_{max} and the radial resolution N_r

$$a^{N_r} = \frac{\rho_{max}}{\rho_0} \quad \text{or} \quad a = e^{1/N_r \ln(\rho_{max}/\rho_0)}. \tag{3}$$

From now on we will use only η ranging from 0 to 2π for the angular component
of the motion field vector.

The mapping of the gray-value function $I(x, y)$ in the cartesian plane to the gray-value function $J(\xi, \eta)$ in the log-plane is by no means trivial. This issue concerns the software implementation of the log-polar mapping given a cartesian image. Every log-polar cell corresponds to a receptive field in the cartesian plane. The image $J(\xi, \eta)$ is the result of a space-variant filtering that affects all subsequent computations on the log-polar plane like spatiotemporal filtering appearing later in this paper. We will not delve in this issue here. It has been extensively studied in [3] but it still remains an open problem. In our implementation we used non-overlapping averaging receptive fields as implemented in the emulation of the space-variant sensor in [22].

3. The Motion Field in the Log-Polar Plane

We begin this section with a brief summary of the cartesian motion equations. Let an object be moving with translational velocity $v = (v_x, v_y, v_z)^T$ and angular velocity $\omega = (\omega_x, \omega_y, \omega_z)^T$ relative to the camera. We denote by X the position of a point on the object with respect to the camera coordinate system, by $\hat{z}$ the unit-vector in the z-axis taken as the optical axis, and by $x = (x, y, 1)^T$ the projection of X on the image plane $Z = 1$. The motion field vector reads [15]

$$\dot{x} = \frac{1}{\hat{z}^T X} \hat{z} \times (v \times x) + \hat{z} \times (x \times (x \times \omega)). \tag{4}$$

In case of ego-motion of the camera with the above velocities and a stationary environment, the above equation as well as all following equations have to be read with the opposite sign for v and ω. The 3D-motion estimation problem is known as the recovery of all the depths, the direction of translation v, and the angular velocity from the motion field $\dot{x}$. The magnitude of translation cannot be recovered due to the scale ambiguity that couples translation and depth.

We first compute the motion field vectors in the polar plane. The definition of the polar coordinates yields

$$\dot{\rho} = \dot{x} \cos \eta + \dot{y} \sin \eta$$

$$\dot{\eta} = \frac{1}{\rho}(-\dot{x} \sin \eta + \dot{y} \cos \eta). \tag{5}$$

The radial component of the log-polar motion field can be easily obtained:

$$\dot{\xi} = \frac{1}{\ln a} \frac{\dot{\rho}}{\rho}. \tag{6}$$

In order to make the equations for the log-polar plane more readable we introduce the polar unit-vectors $\hat{\rho} = (\cos \eta, \sin \eta)^T$ and $\hat{\eta} = (-\sin \eta, \cos \eta)^T$. Furthermore we introduce the vectors $v_{xy} = (v_x, v_y)^T$ and $\omega_{xy} = (\omega_x, \omega_y)^T$ to describe motion parallel to the image plane. We will carry out the computations for the polar motion field $(\dot{\rho}, \dot{\eta})$. The log-polar motion field is different only in the radial component $\dot{\xi}$ which can be computed from $\dot{\rho}$ straightforward (6). This will also enable us to find out that most of the advantages are due to the polar nature of log-polar plane.

Using the cartesian motion field (4) and the transformation rules (5) we obtain the following expressions which relate the polar motion field to the 3D geometry (depths Z) and motion (v, ω) of the scene:

$$\dot{\rho} = -\frac{\rho v_z}{Z} + \frac{v_{xy}^T \hat{\rho}}{Z} + (1 + \rho^2)\omega_{xy}^T \hat{\eta}$$

$$\dot{\eta} = \frac{v_{xy}^T \hat{\eta}}{\rho Z} - \frac{\omega_{xy}^T \hat{\rho}}{\rho} + \omega_z. \tag{7}$$

Hence, the log-component of the motion field reads

$$\ln a \, \dot{\xi} = -\frac{v_z}{Z} + \frac{v_{xy}^T \hat{\rho}}{\rho Z} + \left(\frac{1}{\rho} + \rho\right)\omega_{xy}^T \hat{\eta}. \tag{8}$$

Equivalent relations have been derived in [23].

We see that the motion field is the sum of the translational part including the depth information and a rotational part like in the cartesian formulation. However, we note that the influence of the motion-components parallel to the optical axis are decoupled. The translation v_z appears only in $\dot{\rho}$ and the rotation ω_z only in $\dot{\eta}$. This was expected due to the properties of the complex logarithmic mapping regarding two-dimensional expansions and rotations, respectively. However, as it is already known the main problem in recovery of 3D-motion is to totally decouple the rotational from the translational effects. We will show later how to attack this problem globally by an active technique, and locally by a purposive technique.

3.1. Pure Translation

If there is only translation parallel to the optical axis (Fig. 2a) the polar (and the log-polar) motion field has only one component as can be seen in Fig. 2b. The magnitude variation of the log-component $\dot{\xi}$ in this case depends only on the depths of the projected 3D structures. If the motion is pure translation in an arbitrary direction (Fig. 2c) the motion field looks like in Fig. 2d.

To better understand this field structure and the method presented next we show a pencil of lines in the cartesian plane and its complex logarithmic mapping in Fig. 3. The lines are the orbits of the motion field.

Closer inspection of the equations of the polar translational motion field

$$\dot{\rho} = -\frac{\rho v_z}{Z} \tag{9}$$

$$\dot{\eta} = \frac{v_{xy}^T \hat{\eta}}{\rho Z} \tag{10}$$

Figure 2a–d. Two translational motion fields in the cartesian plane (left) and the log-polar plane (right). The horizontal axis is the η-axis and the vertical axis is the ξ-axis. The η angle is measured counter-clockwise beginning at $-90°$

yields following facts:

1. The angular component reads $\dot{\eta} = v_{xy}^T \eta / \rho Z$. There are two lines in the motion field $\eta = \phi$ and $\eta = \phi + \pi$ (or $\phi - \pi$ depending on which of both is in $[0, 2\pi)$) where $\dot{\eta}$ vanishes for every ρ. The angle ϕ gives the direction of (v_x, v_y) or the line where the focus of expansion lies. Assuming that there is no direction η where the integral

$$\int_{\rho_0}^{\rho_{max}} \frac{1}{\rho^2 Z^2} d\rho$$

vanishes the desired directions ϕ, and $\phi \pm \pi$ are given by the global minima of $\int \dot{\eta}^2 d\rho$ in the presence of noisy measurements. The above sufficient condition is

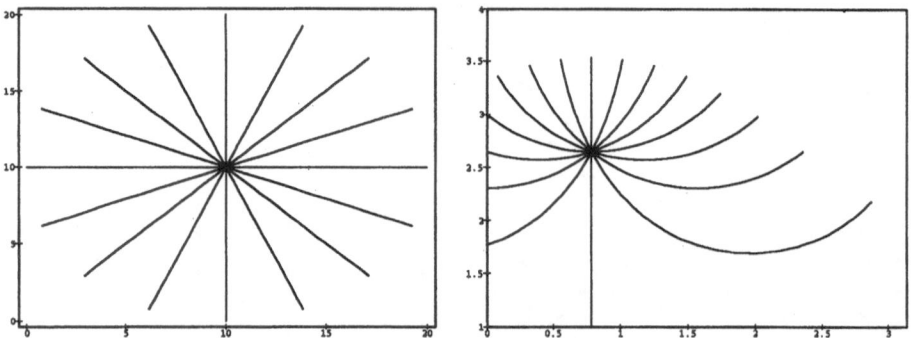

Figure 3. The complex logarithmic mapping of a line pencil. The line through the origin is mapped into a vertical line. The horizontal axis is the η-axis and the vertical axis is the ξ-axis

met if the camera does not gaze on a point of infinite depth. For example, the condition is not satisfied if the origin of the cartesian image lies above the horizon's projection. If the integral $\int \dot{\eta}^2 \, d\rho$ vanishes everywhere we imply the existence of a pure v_z-translation.

2. Along the lines $\eta = \phi + \pi/2$ and $\eta = \phi - \pi/2$ or $(\phi + 3\pi/2)$ the flow reads

$$\dot{\rho}_{\phi+\pi/2} = -\rho\frac{v_z}{Z} \qquad \dot{\eta}_{\phi+\pi/2} = \pm\frac{\sqrt{v_x^2 + v_y^2}}{\rho Z}.$$

By dividing we obtain

$$\frac{\dot{\rho}_{\phi+\pi/2}}{\rho^2 \dot{\eta}_{\phi+\pi/2}} = \pm\frac{v_z}{\sqrt{v_x^2 + v_y^2}} \tag{11}$$

which is the tangent of the polar angle of the translation direction.

Hence, it is possible by an explicit search for the global minima along the angle coordinate—which is feasible in the polar plane due to the low resolution—to obtain the full translation direction.

As an alternative method, we could proceed as in the cartesian case where we first eliminate the depth from the motion field equation (4)—recall that Eq. (4) consists of two equations since $\dot{x}$ has two components. If we eliminate depth from (7) we obtain

$$\rho\dot{\eta}(v_{xy}^T \hat{\rho} - \rho v_z) = \dot{\rho}v_{xy}^T \hat{\eta}$$

that can be rewritten as

$$(v_x \quad v_y \quad v_z) \begin{bmatrix} \rho\dot{\eta}\cos\eta + \dot{\rho}\sin\eta \\ \rho\dot{\eta}\sin\eta - \dot{\rho}\cos\eta \\ -\rho^2\dot{\eta} \end{bmatrix} = 0. \tag{12}$$

This is nothing else than the equivalent of the well known epipolar constraint in the continuous case [5]:

$$v^T(x \times \dot{x}) = 0.$$

3.2. *Fixation in the Log-Polar Plane*

We assume that the camera mount has two controllable degrees of freedom which enable a rotation about an axis parallel to the image plane through the optical center. We denote by $(\Omega_x, \Omega_y, 0)$ the resulting additional angular velocity. The motion field arising from the relative motion of the camera to the environment (ego and/or object motion) follows from (4)

$$\dot{x} = \frac{v_x - xv_z}{Z} - xy(\omega_x - \Omega_x) + (1 + x^2)(\omega_y - \Omega_y) - y\omega_z$$

$$\dot{y} = \frac{v_y - yv_z}{Z} - (1 + y^2)(\omega_x - \Omega_x) + xy(\omega_y - \Omega_y) + x\omega_z. \tag{13}$$

We define fixation as keeping the same point in the image center along time. The fixated point may be a point on a moving object in case of a stationary or moving observer or a stationary point in the case of a moving observer. Fixation is achieved by closed loop control where controllables are the pan and the tilt angle of the camera. The formally defined fixation criterion we use is the vanishing of the motion field vector at the central point of the image:

$$\dot{x}|_{x=0} = 0. \tag{14}$$

This is equivalent to

$$\frac{v_x}{Z_o} + \omega_y - \Omega_y = 0 \quad \text{and} \quad \frac{v_y}{Z_o} - (\omega_x - \Omega_x) = 0, \tag{15}$$

where Z_o is the depth of the point projected on the image center.

However, this criterion relies on the existence of a well defined mathematical point in the image. In reality, this is only the case when the camera fixates on an object like a light spot. When fixating on a larger object the average flow over an area should be minimized. Assuming that this area is the cartesian fovea we seek control angular velocities (Ω_x, Ω_y) such that

$$\int_{r=0..\rho_0} \int_{\theta=0..2\pi} \dot{x} r \, dr \, d\theta = 0$$

where $\dot{x} = (\dot{x}, \dot{y})$ is given in (13) and ρ_0 is the foveal radius. We note that we do not minimize the integral of the squares of the flow so that we allow foveal motions like rolling around the optical axis. After integration we obtain

$$\Omega_y = \omega_y + v_x \frac{\zeta}{1 + \frac{\rho_0^2}{4}} \quad \text{and} \quad \Omega_x = -\omega_x + v_y \frac{\zeta}{1 + \frac{\rho_0^2}{4}} \tag{16}$$

where ζ is the average of the inverse depth. For a small foveal area we can still use the approximation (15).

What kind of control strategy is applied to achieve the fixation criterion is not the subject of this paper. We will assume that the system is in steady state mode and we will study the effects of the satisfied fixation criterion in the peripheral motion-field of the log-polar plane.

We turn now to the motion field on the polar transform of the periphery which reads in case of fixation:

$$\dot{\rho} = -\frac{\rho v_z}{Z} + \frac{v_{xy}^T \hat{\rho}}{Z} + (1 + \rho^2)(\omega_{xy} - \Omega_{xy})^T \hat{\eta}$$

$$\dot{\eta} = \frac{v_{xy}^T \hat{\eta}}{\rho Z} - \frac{(\omega_{xy} - \Omega_{xy})^T \hat{\rho}}{\rho} + \omega_z . \tag{17}$$

We rewrite the polar motion field by means of (15)

$$\dot{\rho} = -\frac{\rho v_z}{Z} + \frac{v_{xy}^T \hat{\rho}}{Z} + (1 + \rho^2)\left(\frac{v_y}{Z_o} \quad -\frac{v_x}{Z_o}\right)\hat{\eta}$$

$$\dot{\eta} = \frac{v_{xy}^T \hat{\eta}}{\rho Z} - \frac{1}{\rho}\left(\frac{v_y}{Z_o} \quad -\frac{v_x}{Z_o}\right)\hat{\rho} + \omega_z . \tag{18}$$

that leads to the following polar motion field in case of fixation:

$$\dot{\rho} = -\frac{\rho v_z}{Z} + \left(\frac{1}{Z} - \frac{1 + \rho^2}{Z_o}\right)v_{xy}^T \hat{\rho}$$

$$\dot{\eta} = \frac{1}{\rho}\left(\frac{1}{Z} - \frac{1}{Z_o}\right)v_{xy}^T \hat{\eta} + \omega_z . \tag{19}$$

We rewrite the equation by introducing $v_z' = v_z/Z_o$ and $v_{xy}' = v_{xy}/Z_o$:

$$\dot{\rho} = -\frac{Z_o}{Z}\rho v_z' + \left(\frac{Z_o}{Z} - (1 + \rho^2)\right)v_{xy}'^T \hat{\rho}$$

$$\dot{\eta} = \frac{1}{\rho}\left(\frac{Z_o}{Z} - 1\right)v_{xy}'^T \hat{\eta} + \omega_z . \tag{20}$$

To reduce the number of symbols we will use again non-primed symbols for the translation. We first note that we obtain as independent unknowns the depths Z_o/Z relative to the depth of the fixation point which results in an object centered scene representation. Second, the other unknowns are reduced from initially five in the non-fixation case—three for rotation and two for the translation direction—to four: the three scaled translation components, and one angular velocity component. This dimension reduction was already proved in the cartesian plane in [1] and [2].

We next propose an algorithm to recover the motion parameters in case of fixation from the motion field in the log-polar plane. Equivalent to the case of pure transla-tion without fixation our method is based on the observation that the angular component $\dot{\eta}$ equals ω_z everywhere along the line $\eta = \phi$ where $\tan\phi = v_y/v_x$. However, this is not a necessary condition. The angular component $\dot{\eta}$ equals ω_z in the additional case that $Z = Z_0$ along a radial line, or more general $\dot{\eta}$ is constant if the relative depth is linearly varying:

$$\frac{Z_o}{Z} - 1 = k\rho. \tag{21}$$

This may happen if the environment is planar. For a particular η_0 the radial component for $\eta = \eta_0$ reads in this case

$$\dot{\rho} = -(1 + k\rho)v_z + (k\rho - \rho)^2 v_{xy}^T \hat{p}_{\eta_0}.$$

We can exclude this case if we test $\dot{\rho}$ subject to quadratic variation variation with respect to ρ. The algorithm we propose comprises following steps:

1. We build the average $\bar{\dot{\eta}}$ over ρ for every η and the variance

$$\int_\rho (\dot{\eta} - \bar{\dot{\eta}})^2 \, d\rho.$$

The variance vanishes at $\eta = \phi$ and $\eta = \phi \pm \pi$ and at the angles where $Z = Z_0$ on an entire radial line which is excluded as described above.

We carry out an one-dimensional search for the global minima of the above integral. We carry out the above described linearity test to exclude the case $Z = Z_0$. The global minima give the direction of the line containing the FOE in the cartesian plane as well as ω_z (equal the average $\bar{\dot{\eta}}$ at $\eta = \phi$). We exclude additional global minima due to linear depth variation as described above.

2. At $\eta = \phi + \pi/2$ and $\eta = \phi - \pi/2$ (or $\phi + 3\pi/2$) the motion field reads:

$$\dot{\rho} = -\frac{Z_o}{Z}\rho v_z$$

$$\dot{\eta} = \frac{1}{\rho}\left(\frac{Z_o}{Z} - 1\right)\sqrt{v_x^2 + v_y^2} + \omega_z.$$

We build the following averages

$$\overline{\rho\dot{\eta}} = \bar{\rho}\omega_z + (\overline{Z_o/Z} - 1)\sqrt{v_x^2 + v_y^2}$$

$$\overline{\dot{\rho}/\rho} = -\overline{Z_o/Z}v_z.$$

3. Building for every point the deviation from the average we obtain

$$\frac{\dot{\rho}}{\rho} - \overline{\dot{\rho}/\rho} = -v_z\left(\frac{Z_o}{Z} - \overline{Z_o/Z}\right)$$

$$\rho\dot{\eta} - \overline{\rho\dot{\eta}} - (\rho - \bar{\rho})\omega_z = \sqrt{v_x^2 + v_y^2}\left(\frac{Z_o}{Z} - \overline{Z_o/Z}\right).$$

After testing the vanishing of the lhs of the latter equation indicating a frontal translation we obtain

$$\frac{v_z}{\sqrt{v_x^2 + v_y^2}} = -\frac{\dfrac{\dot{\rho}}{\rho} - \overline{\dot{\rho}/\rho}}{\rho\dot{\eta} - \overline{\rho/\dot{\eta}} - (\rho - \bar{\rho})\omega_z} \tag{22}$$

for every point along the lines $\eta = \phi + \pi/2$ and $\eta = \phi - \pi/2$ (or $\phi + 3\pi/2$). Since ω_z is given by the second step we are able to recover the angle between translation and the optical axis. Its tangent is equal to the inverse of the above expression.

We elaborated a method to recover the direction of translation in the case of fixation from the polar motion field based on a feasible 1D search over the angle range. The equations for the logarithmic field are obtained straightforward by substituting (6) in the above expressions. This substitution does not introduce any computational advantages, hence, the potential of the method relies only on the polar nature of the log-polar mapping.

3.3. Experiments on the Translation Computation

In this subsection, we present results on the computation of the optical flow in the log-polar domain as well as on the computation of the translation direction in the case of pure translation.

We tested the method proposed in Section 3 in a log-polar sequence obtained from the cartesian real world sequence "Marbled Block"[1] [16]. One image of the sequence and its log-polar transform are shown in Fig. 4 top and middle, respectively. The log-polar image is drawn such that the η-axis is the horizontal axis and the ξ-axis is the vertical axis pointing downwards. To interpret the log-polar images we note that the angle η is measured beginning counterclockwise from the y-axis which is pointing downwards. So moving horizontally in the log-polar plane we first see the transformed lower right quadrant, then the transformed upper right quadrant and so on. The compression rate obtained by the log-polar transformation is $1:25$.

The optical flow is computed using the spatiotemporal derivatives and the assumption that the flow is locally affine in the cartesian domain. Based on the computed log-polar flow of the "Marbled Block" sequence we apply the method in Section 3 to estimate the direction of translation. We show in Fig. 5 the computed average angular component $\dot{\eta}$ as a function of the angle. We obtain two global minima— differing by 180 degrees as expected—giving as result an angle of 253 degrees for the direction of the line containing the FOE. By averaging the expression

$$\frac{\dot{\rho}_{\phi+\pi/2}}{\rho^2\dot{\eta}_{\phi+\pi/2}}$$

---

[1] Created by Michael Otte at University of Karlsruhe and FhG-IITB, Germany.

Figure 4. The original cartesian image (top), the log-polar transformed image (middle) and the computed optical flow in the log-polar domain (bottom) for the Marbled Block sequence

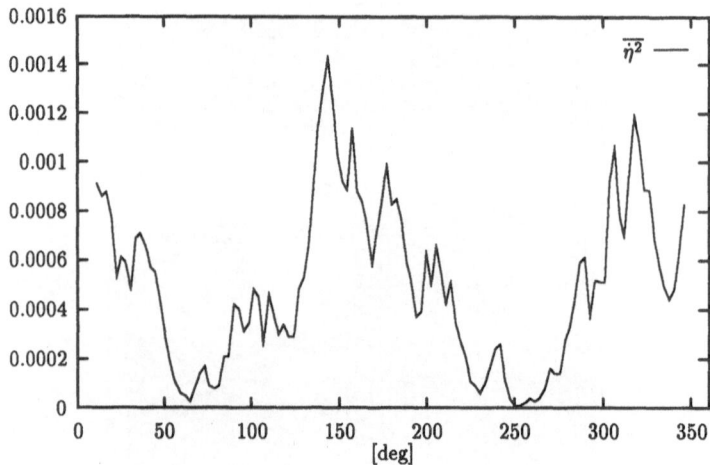

Figure 5. The average of the squared angular components along the radius as a function of the angle η for the Marbled Block sequence

over the diameter line corresponding to $253 - 90 = 163$ degrees we obtain as estimated cartesian FOE-position $(-247, 102)$. Using the intrinsic calibration parameters of the sequence we find out that the angle error between the veridical and the estimated translation direction is 5 degrees which is an acceptable estimation error given the fact that the computational effort decreased to its 4%.

4. Local Motion Computations

In the prior sections we presented methods which exploited the log-polar motion field over the entire peripheral field of view—the domain of log-polar mapping. Here, we delve into the variation of the motion field—also called motion parallax —in the immediate neighborhood of every point. It was repeatedly proved [7, 11, 20, 27] that the differential invariants of the optical flow divergence, curl, and the two shear components contain information on motion and depth.

In this section, we will study the local motion parallax in the log-polar plane. We start with a brief description of the parallax in the cartesian plane. After differentiating the motion field $\dot{x}$ in (4) with respect to x and y we obtain the field divergence

$$ \text{div} = \frac{\partial \dot{x}}{\partial x} + \frac{\partial \dot{y}}{\partial y} = \frac{-2v_z - (pv_x + qv_y) + 3v_z(px + qy)}{Z(1 - px - qy)} - 3y\omega_x + 3x\omega_y, \quad (23) $$

where

$$ p = \frac{\partial Z}{\partial X} \qquad q = \frac{\partial Z}{\partial Y}, $$

and $(-p, -q, -1)$ is the normal to the tangential plane of the viewed surface. The shear components are similarly obtained

$$\text{shear}_1 = \frac{\partial \dot{x}}{\partial x} - \frac{\partial \dot{y}}{\partial y} = \frac{-p(v_x - xv_z) + q(v_y - yv_z)}{Z(1 - px - qy)} + y\omega_x + x\omega_y$$

$$\text{shear}_2 = \frac{\partial \dot{x}}{\partial y} + \frac{\partial \dot{y}}{\partial x} = \frac{\partial \dot{x}}{\partial x} - \frac{\partial \dot{y}}{\partial y} = \frac{-q(v_x - xv_z) - p(v_y - yv_z)}{Z(1 - px - qy)} - x\omega_x + y\omega_y \qquad (24)$$

We do not consider the curl of the motion field because it contains information about the ω_z component which is not of interest to the behavioral tasks we study. The above expressions are simplified at the center of the image ($x = 0, y = 0$) where they contain only translational and depth information. Furthermore if the translation is parallel to the optical axis ($v_x = 0, v_y = 0$) the divergence equals v_z/Z which is the inverse of the time to collision (TTC) with the object viewed in the image center. If the translation is arbitrary Subbarao [20] proved that the inverse of the TTC can be bounded by the two values div $\pm \|shear\|$. Unfortunately, if points outside the fovea are considered—as is the case in the log-polar motion field—the above parallax expressions contain rotational information that cannot be eliminated. Parallax methods have either considered the rotational terms as negligible [4] or have used a different image surface. Indeed, if the scene is projected on a spherical surface, the divergence and shear components of the resulting motion field—which is tangential to the image sphere—get rid of the rotation information [11]. This is based on the fact that the rotational *spherical* motion field is linear in (x, y) whereas the rotational *planar* motion field contains terms quadratic in the image coordinates.

Although there are many practical situations where the cartesian invariants could yield acceptable results for the time to collision and the surface slant—the examples from the literature are very rare—we will search here for exact relations. The inspiration of this study was the result in [23] we will begin with.

We denote the derivatives of depth with respect to ρ and η with Z_ρ and Z_η. We first compute the following spatial derivatives:

$$\frac{\partial \dot{\eta}}{\partial \eta} = -\frac{Z_\eta}{\rho Z^2} v_{xy}^T \hat{\eta} - \frac{1}{\rho Z} v_{xy}^T \hat{\rho} - \frac{1}{\rho} \omega_{xy}^T \hat{\eta} \qquad (25)$$

$$\frac{\partial \dot{\rho}}{\partial \rho} = \frac{\rho Z_\rho - Z}{Z^2} v_z - \frac{Z_\rho}{Z^2} v_{xy}^T \hat{\rho} + 2\rho \omega_{xy}^T \hat{\eta}. \qquad (26)$$

In combination with the scaled field component

$$\frac{\dot{\rho}}{\rho} = -\frac{v_z}{Z} + \frac{1}{\rho Z} v_{xy}^T \hat{\rho} + \left(\frac{1}{\rho} + \rho\right) \omega_{xy}^T \hat{\eta} \qquad (27)$$

We see that the only linear combination of the above three expressions that is independent of rotation is

$$2\frac{\partial \dot{\eta}}{\partial \eta} - \frac{\partial \dot{\eta}}{\partial \rho} + 2\frac{\dot{\rho}}{\rho} = -\frac{v_z}{Z} + \frac{Z_\rho}{Z^2}(v_{xy}^T \hat{\rho} - \rho v_z) - 2\frac{Z_\eta}{\rho Z^2}v_{xy}^T \hat{\eta}. \qquad (28)$$

If we assume that the surface is locally frontal then the depth derivatives vanish [23] and we obtain the inverse of the time to collision v_z/Z. However, this assumption is not always valid. In high eccentricities the pixels of the log-polar plane span a considerable viewing angle in space that may contain a non negligible depth variation. Therefore, we compute the depth derivatives Z_ρ and Z_η as a function of the normal of the surface. Suppose that the surface normal is $(-p, -q, -1)$ with $p = \partial Z/\partial X$ and $q = \partial Z/\partial Y$. It can be easily proved that

$$\frac{\partial Z}{\partial x} = \frac{\rho Z}{1 - px - qy} \qquad \frac{\partial Z}{\partial y} = \frac{\partial Z}{1 - px - qy}$$

Let us denote (p, q) with p. The required derivatives read

$$Z_\rho = \frac{Z p^T \hat{\rho}}{1 - \rho p^T \hat{\rho}} \quad \text{and} \quad Z_\eta = \frac{Z \rho p^T \hat{\eta}}{1 - \rho p^T \hat{\rho}}. \qquad (29)$$

Hence,

$$2\frac{\partial \dot{\eta}}{\partial \eta} - \frac{\partial \dot{\rho}}{\partial \rho} + 2\frac{\dot{\rho}}{\rho} = \frac{-v_z + v_{xy}^T \hat{\rho} p^T \hat{\rho} - 2v_{xy}^T \hat{\eta} p^T \hat{\eta}}{Z(1 - \rho p^T \hat{\rho})}. \qquad (30)$$

The exact expression has the advantage of being rotation independent but depends on the unknown normal of the tangential plane of the viewed surface. We know [20] that we can only *bound* the time to collision by using both divergence and shear. However, in a planar image this is only possible in the *rotation-free* case. Using a spherical image the bounds become rotation-independent because the dependence of the motion field on rotation becomes linear $\omega \times x$ (cf. the quadratic rotational terms in (4) and (5)). Earlier approaches have assumed that near the center a planar image can be approximated by a spherical image [7]. However, this approximation is not valid in the non-foveal part considered in the log-polar mapping. We will avoid this approximation by transforming exactly onto the polar plane.

Let us assume that the image is the unit half-sphere and that the position on this sphere is given by the polar angle θ and the azimuth angle ϕ. Regarding the polar coordinate system above, we have

$$\rho = \tan \theta \quad \text{and} \quad \eta = \phi.$$

The velocity $\dot{s}$ of a point s on the image sphere is on the tangential plane at that point and is given by

$$\dot{s} = \dot{\theta}\hat{\theta} + \dot{\phi}\sin\theta\hat{\phi} \qquad (31)$$

where $\hat{\theta}$ and $\hat{\phi}$ are the unit-vectors of the spherical coordinate system respectively. In analogy to the differential invariants on a spherical image we next propose parallax expressions in the polar motion field that provide bounds for the time to collision which are independent both from the surface slant and the rotation. The divergence of the spherical motion field is [12]

$$\operatorname{div} \dot{s} = \frac{\partial \dot{\theta}}{\partial \theta} + \frac{\partial \dot{\phi}}{\partial \phi} + \frac{\dot{\theta}}{\tan \theta} \tag{32}$$

and the two shear components read [11]

$$\operatorname{shear}_1 \dot{s} = \frac{\partial \dot{\theta}}{\partial \theta} - \frac{\partial \dot{\phi}}{\partial \phi} - \frac{\dot{\theta}}{\tan \theta} \quad \text{and} \quad \operatorname{shear}_2 \dot{s} = \frac{1}{\sin \theta} \frac{\partial \dot{\theta}}{\partial \phi} + \sin \theta \frac{\partial \dot{\phi}}{\partial \theta}. \tag{33}$$

By rewriting the right hand sides of the above equations using a polar coordinate system ($\rho = \tan \theta, \eta = \phi$) on the image plane we obtain

$$\operatorname{div} \dot{s} = \frac{\partial \dot{\rho}}{\partial \rho} + \frac{\partial \dot{\eta}}{\partial \eta} + \frac{\dot{\rho}}{1 + \rho^2}\left(\frac{1}{\rho} - 2\rho\right)$$

$$\operatorname{shear}_1 \dot{s} = \frac{\partial \dot{\rho}}{\partial \rho} - \frac{\partial \dot{\eta}}{\partial \eta} - \frac{\dot{\rho}}{1 + \rho^2}\left(\frac{1}{\rho} + 2\rho\right)$$

$$\operatorname{shear}_2 \dot{s} = \frac{\sqrt{1 + \rho^2}}{\rho}\left((1 + \rho^2)\frac{\partial \dot{\eta}}{\partial \rho} + \frac{1}{1 + \rho^2}\frac{\partial \dot{\rho}}{\partial \eta}\right). \tag{34}$$

Then we build the squared shear magnitude

$$\operatorname{shear}^2 \dot{s} = \operatorname{shear}_1^2 \dot{s} + \operatorname{shear}_2^2 \dot{s}$$

and after tedious calculations we find out that it is factorizable:

$$\operatorname{shear}_1 \dot{s}^2 + \operatorname{shear}_2 \dot{s}^2 = \|S_\alpha\|^2 \|S_\beta\|^2 \tag{35}$$

with

$$S_\alpha = \frac{1}{Z}\begin{bmatrix} Z_\rho(1 + \rho^2) + \rho \\ Z_\eta\sqrt{1 + \rho^2} \\ Z\rho \end{bmatrix} \qquad S_\beta = \frac{1}{Z}\begin{bmatrix} \dfrac{v_{xy}^T \hat{\rho} - \rho v_z}{1 + \rho^2} \\ \dfrac{v_{xy}^T \hat{\eta}}{\sqrt{1 + \rho^2}} \end{bmatrix}.$$

On the other hand, trying to express the divergence as a function of S_α and S_β we find out that

$$\operatorname{div} \dot{s} = -\frac{2(\rho v_{xy}^T \hat{\rho} + v_z)}{Z(1 + \rho^2)} - S_\alpha^T S_\beta. \tag{36}$$

Since $S_\alpha^T S_\beta \le \|S_\alpha\| \|S_\beta\|$ we obtain

$$\operatorname{div} - \|\operatorname{shear}\| < -\frac{2(\rho v_{xy}^T \hat{\rho} + v_z)}{Z(1 + \rho^2)} \le \operatorname{div} + \|\operatorname{shear}\|. \tag{37}$$

The expression in the middle can be rewritten as

$$\frac{(\rho v_{xy}^T \hat{\rho} + v_z)}{Z(1 + \rho^2)} = \frac{v^T \hat{x}}{\|X\|}$$

where $\hat{x}$ is the unit-vector in the space direction of the point $x = (x, y, 1)$ and $\|X\|$ is the distance to the 3D-point X. This expression is referred in the literature as the

K. Daniilidis

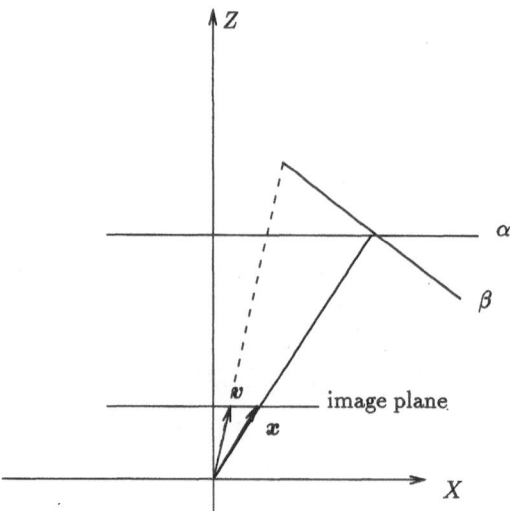

Figure 6. Geometric interpretation of different definitions for the time to collision. The time to collision Z/v_z is the time required for the observer to hit the plane α at depth Z or the time required by an object like a frontoparallel plane α to hit the observer. The second definition $v^T \hat{x}/\|X\|$ is the time required by the observer to hit the plane β perpendicular to the line of sight at the considered point

inverse of the time to collision [7] or looming [17]. However, it differs from the more frequently definition v_z/Z. The former deals with distances $\|X\|$ to the points, whereas the latter deals with the depths Z of the points. The physical interpretation is shown in Fig. 6. The time to collision Z/v_z is the time required for the observer to hit the plane α at depth Z or the time required by an object like a frontoparallel plane α to hit the observer. It is always geometrically plausible in the sense of a scaled depth map but it is physically plausible only in case of frontal translation. On the other hand, the looming expression $v^T \hat{x}/\|X\|$ is the time required by the observer to hit the plane β perpendicular to the line of sight at the considered point. It is neither a scaled depth nor a scaled range but it can be interpreted as the inverse of the time to collision if the line of sight is in the translation direction. Furthermore, it achieves its maximum at this point—the focus of expansion—if we assume that all points are equidistant $\|X\| = constant$.

In contrast first to the rotation dependent cartesian bounds of v_z/Z, and second to the slant dependent expression in polar coordinates in (30), we derived both rotation and slant independent bounds on a new expression for the time to collision in the polar plane. All the expressions appearing in (34) may be easily transformed from the polar to the logpolar plane using

$$\dot{\rho} = \ln a \rho_0 a^\xi \dot{\xi} \qquad \frac{\partial \dot{\rho}}{\partial \rho} = \ln a \; \dot{\xi} + \frac{\partial \dot{\xi}}{\partial \xi}$$

$$\frac{\partial \dot{\rho}}{\partial \eta} = \ln a \rho_0 a^\xi \frac{\partial \dot{\xi}}{\partial \eta} \qquad \frac{\partial \dot{\eta}}{\partial \rho} = \frac{\rho_0 a^\xi}{\ln a} \frac{\partial \dot{\eta}}{\partial \xi}. \tag{38}$$

Despite the significant data reduction no further advantages are obvious from the transition from polar to log-polar coordinates.

5. Conclusion

An essential feature of an attentive visual system is space-variant sensing. In this study we used the complex logarithmic mapping to model the resolution decrease in the image periphery. The achieved compression ratio of up to $1:25$ enables reactive behavior in real time.

We introduced two new methods for computing the focus of expansion by exploiting the structure of the flow patterns in the log-polar motion field. Both are based on a 1D global minimum search which is inexpensive due to the low angular resolution. The second method necessitates the pursuit movement of the camera proving, thus, that a full attentional mechanism based on both fixation and space-variant sensing enables motion estimation in case of general motion. We implemented the first method in a translating real word sequence obtaining an estimation error of $5°$ in the translation direction. We plan to implement the second method in the near future using the controllable degrees of freedom of an active binocular head. The second part of our study was devoted to derive motion parallax expressions which yield the time to collision. Based on the parallax equations on a spherical image we derived exact equations for the transformation onto the polar plane. Then, we recovered bounds on the time to collision that are independent of rotation and the surface slant at every position of the image periphery. The mathematical advantages of our approach are due to the polar representation in the log-polar image whereas the complexity advantages are due to the logarithmic nature of the log-polar image.

Acknowledgements

The Vision Group at the DIST Laboratory, University of Genova provided the software emulation of the log-polar sensor. I am grateful to Volker Krüger for the inspiring discussions and his contribution to the software implementation of the flow algorithms. I am indebted to Joannis Papavassiliou for his helpful comments and his hospitality.

References

[1] Aloimonos, J., Weiss, I., Bandyopadhyay, A.: Active vision. Int. J. Comput. Vision *1*, 333–356 (1988).
[2] Bandopadhay, A., Ballard, D. H.: Egomotion perception using visual tracking. Comput. Intell. 7, 39–47 (1990).
[3] Bolduc, M., Levine, M. D.: A foveated retina system for robotic vision. In: ECCV-94 Workshop on Natural and Artificial Visual Sensors, 1994.
[4] Bouthemy, P., Francois, E.: Motion segmentation and qualitative dynamic scene analysis from a moving camera. Int. J. Comput. Vision *10*, 157–182 (1993).
[5] Bruss, A., Horn, B. K. P.: Passive navigation. Comput. Vision Graphics Image Proc. *21*, 3–20 (1983).

[6] Carpenter, R. H. S.: Movements of the eyes. London: Pion Press 1988.

[7] Cipolla, R., Blake, A.: Surface orientation and time to contact from image divergence and deformation. In: Proc. Second European Conference on Computer Vision, pp. 187–202. Santa Margerita, Italy, May 23–26, 1992.

[8] Fermuller, C., Aloimonos, Y.: The role of fixation in visual motion analysis. Int. J. Comput. Vision *11*, 165–186 (1993).

[9] Guissin, R., Ullman, S.: Direct computation of the focus of expansion from velocity field measurements. In: Proc. IEEE Workshop on Visual Motion, pp. 146–155, Princeton, NJ, Oct. 7–9, 1991.

[10] Jain, R., Bartlett, S. L., O'Brien, N.: Motion Stereo Using Ego-Motion Complex Logarithmic Mapping. IEEE Trans. Pattern Anal. Machine Intell. *9*, 356–369 (1987).

[11] Koenderink, J. J., van Doorn, A. J.: Local structure of movement parallax of the plane. J. Opt. Soc. Am. *66*, 717–723 (1976).

[12] Korn, G. A., Korn, T. M.: Mathematical handbook for scientists and engineers. New York: McGraw-Hill 1968.

[13] Mallot, H. A., von Seelen, W., Giannakopoulos, F.: Neural mapping and space-variant image processing. Neural Networks *3*, 245–263 (1990).

[14] Massone, L., Sandini, G., Tagliasco, V.: "Form-invariant" topological mapping strategy for 2D shape recognition. Comput. Vision Graphics Image Proc. *30*, 169–188 (1985).

[15] Negahdaripour, S., Horn, B. K. P.: Direct passive navigation. IEEE Trans. Pattern Anal. Machine Intell. *PAMI 9*, 168–176, (1987).

[16] Otte, M., Nagel, H.-H.: Optical flow estimation: advances and comparisons. In: Proc. Third European Conference on Computer Vision, pp. 51–60. Stockholm, Sweden, May 2–6, 1994.

[17] Raviv, D., Herman, M.: Visual serving from 2-D image cues. In: Active perception (Aloimonos, Y., ed.), pp. 191–226. Hillsdale: Lawrence Erlbaum Associates 1993.

[18] Schultze, M.: Zur Anatomie und Physiologie der Retina. Archiv Mikrosk. Anatomie *2*, 175–286 (1866).

[19] Schwartz, E. L.: Spatial mapping in the primate sensory projection: analytic structure and relevance to perception. Biol. Cybernetics *25*, 181–194 (1977).

[20] Subbarao, M.: Bounds on translational and angular velocity components from first order derivatives of image flow. In: Proc. Conf. American Association of Artificial Intelligence, pp. 744–748. Seattle, WA, July 13–17, 1988.

[21] Taalebinezhaad, M. A.: Direct recovery of motion and shape in the general case by fixation. IEEE Trans. Pattern Anal. Machine Intell. *PAMI-14*, 847–853 (1992).

[22] Tistarelli, M., Sandini, G.: Dynamic aspects in active vision. CVGIP: Image Understanding *56*, 108–129 (1992).

[23] Tistarelli, M., Sandini, G.: On the advantages of polar and log-polar mapping for direct estimation of time-to-impact from optical flow. IEEE Trans. Pattern Anal. Machine Intell. *PAMI-15*, 401–410 (1993).

[24] Van der Spiegel, J., Kreider, G., Claeys, C., Debusschre, I., Sandini, G., Dario, P., Fantini, F., Belluti, P., Soncini, G.: A foveated retina-like sensor using CCD technology. In: Analog VLSI and neural systems (Mead, C., ed.), pp. 189–212. Addison-Wesley: Reading 1989.

[25] van Doorn, A. J., Koenderink, J., Bouman, M. A.: The influence of the retinal inhomogeneity on the perception of spatial patterns. Kybernetik *10*, 223–230 (1972).

[26] Wallace, R. S., Ong, P.-W., Bederson, B. B., Schwartz, E. L.: Space variant image processing. Int. J. Comput. Vision *13*, 71–90 (1994).

[27] Waxman, A. M., Kamgar-Parsi, B., Subbarao, M.: Closed-form solutions to image flow equations for 3D structure and motion. Int. J. Comput. Vision *1*, 239–258 (1987).

[28] Weiman, C. F. R., Chaikin, G.: Logarithmic spiral grids for image processing and display. Comput. Graphics Image Proc. *11*, 197–226 (1979).

Dr. K. Daniilidis
Computer Science Institute
Christian-Albrechts-University Kiel
Preusserstrasse 1–9
D-24105 Kiel
Federal Republic of Germany
e-mail: kd@informatik.uni-kiel.de

Computing Suppl. 11, 21–36 (1996)

Invariant Thinning and Distance Transform

U. Eckhardt and L. Latecki, Hamburg

Abstract

Invariant Thinning and Distance Transform. Thinning is a preprocessing method which is applied to binary (i.e. black-and-white) digital (i.e. discretized) images. The goal of thinning is to reduce the sets of black points in the image to "thin" sets while retaining the "topology" of them as well as "form" properties. Usually thinning methods are organized in an iterative way by "peeling off" outer layers of the sets under consideration. This implies that thinning is an extremely time-consuming task. Recently, Neusius and Olszewski [12] proposed a thinning method which is based on a distance transform. This idea is indeed not new (see e.g. [6]), but Neusius and Olszewski were the first to treat it in a systematic way. Since the distance transform can be calculated efficiently by a two-sweep method, this approach looks attractive. The aim of this paper is to show that under certain assumptions a 'classical' thinning method [4], which has invariance with respect to motions as a distinctive feature, also can be interpreted as a distance-transform-based method.

Key words: Parallel thinning, distance transform, medial axis.

1. Introduction

Thinning is a well-established method for preprocessing binary images. The goal of thinning a binary digital image is to reduce the amount of black points while retaining essential properties of the image. The remaining digital set, the so-called *skeleton* of the original objects should therefore posess some important properties to make the process meaningful and efficient. These properties are

- The skeleton should be "thin".
- The skeleton should have the same topological structure as the original point set.
- The skeleton should reflect in some sense the "shape" of the object.

Due to the importance of thinning algorithms, there exists a large amount of literature on them. For example in [7] performance of thinning algorithms is analyzed. In spite of this amount of literature, there are many open problems in this area:

- The continuous concept underlying the idea of thinning is the medial axis of a set. There exists some theory of the medial axis for the Euclidean norm (see e.g. [14, 15]). However, in the usual setting of digital geometry, the adequate norms are polyhedral ones.

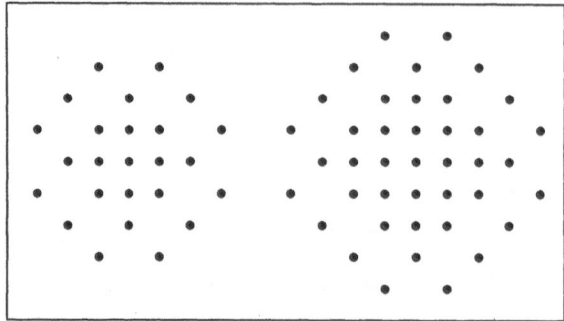

Figure 1. Arcelli-sets

- The medial axis transform is known to be "incorrectly posed" in the Hausdorff metric for sets, which means that small changes of the original set can cause large changes in the medial axis of it (see [16]).
- Almost all thinning methods, whether they are intended for sequential or parallel implementation, proceed in a basically sequential way. This means that they are organized in "phases" in which only a certain subset of boundary points is eliminated. This implies that the result of their application ist not invariant with respect to motions of the digital plane.
- The result of skeletonization should be a "thin" skeleton. Arcelli [1] presented 1981 the first example of a set which is not "thin" in any conceivable sense but nevertheless irreducible, i.e. its own skeleton (see Fig. 1). In [3] a characterization of such thick irreducible sets is given.

The usual approach for thinning is to eliminate iteratively boundary points of the set under consideration. Therefore one gets a method whose efficiency depends on the thickness of the set to be processed. This is the reason for the observation that many authors presented methods which are not iterative in nature but make use of the distance transformation which associates to each point of a set its distance to the complement of the set. When doing so, usually the topological properties are lost, i.e. Euler's number of the given set is changed. For example, the set given in Figure 8a is connected, but its skeleton obtained by a distance transformation is not connected (Figure 8c). Neusius and Olszewski [12] proposed a distance-transformation based method with the additional property that the result of its application does not change the number of connected components of the set and of its complement.

In this paper a different approach is investigated. A certain known thinning method is shown to be distance-transform based if some conditions are fulfilled. It is an interesting fact that the condition of "well-composedness" which was introduced elsewhere [8] plays an important role also in this context.

2. Basic Definitions

The *digital plane* $\mathbb{Z}^2$ is the set of all points in the Euclidean plane having integer coordinates. A *digital set* is a subset of the digital plane.

Each point P in the digital plane has eight *neighbors*, which are numbered 0 to 7 according to the following scheme

$$
\begin{array}{ccc}
N_3(P) & N_2(P) & N_1(P) \\
N_4(P) & P & N_0(P) \\
N_5(P) & N_6(P) & N_7(P)
\end{array}
$$

$\mathcal{N}(P)$ is the set of all neighbors of P (without P itself) and is termed the *neighborhood* of P. For $i \in \mathbb{Z}$ we understand $N_i(P)$ in an obvious way to be the i mod 8-th neighbor in $\mathcal{N}(P)$. Neighbors with even numbers are *direct neighbors* of P or *4-neighbors*; neighbors having odd numbers are *indirect neighbors* of P. $\mathcal{N}_4(P)$ is the set of all direct neighbors of P (without P itself) and is termed the *4-neighborhood*. The neighbors in $\mathcal{N}(P)$ are sometimes called 8-*neighbors*. By means of the neighbor-relation we are able to define a 'topology' in the digital plane [13]. It was also shown by Rosenfeld that this topology has especially favourable properties if the set S is equipped with 8-connectedness relation and its complement with 4-connectedness. In this case one can prove a 'Jordan property' for digital 'curves' [13].

For a point P in a digital set S we define the 8-*connection number* $C_8(P)$ to be the number of 8-connected components in $\mathcal{N}(P) \cap S$. Similarly, the number of 4-connected components in $\mathcal{N}(P) \cap S$ which contain direct neighbors of P is termed the 4-connection number $C_4(P)$ of P in S.

A point of a digital set S having all four direct neighbors also in S is an *interior point*. All points in S which are not interior points are *boundary points*. The set of all interior points of a set S is termed the *kernel* or *interior* of S and is denoted int S. The *inner boundary* of S is the set of all boundary points of int S or equivalently, the set of all interior points having a direct neighbor which is a boundary point of S.

A point in S is termed (4/8-)*simple*, if it is a boundary point and if $C_{4/8}(P) = 1$ [13].

Thinning a digital set means application of an iterative procedure for removing simple points. A set is termed (4/8-)*irreducible* if it does not contain any (4/8-) simple points.

A *perfect configuration* of a digital set consists of a triple of points, P, $N_{2i}(P)$, and $N_{2i+4}(P)$, such that $P \in S$, $N_{2i}(P) \in$ int S and $N_{2i+4}(P) \notin S$. P is then termed a *perfect point* of S and $N_{2i}(P)$ is a *perfect partner* of P. Observe that the perfect partner is always an interior point. For example, if P is black in the following set, then it is perfect with perfect partner $N_4(P)$:

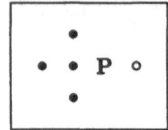

A digital set it is called *p-irreducible* if it does not contain any points which are simple and perfect. When only simple and perfect points are used for thinning, then a thinning method is obtained which is inherently parallel and invariant with respect to motions of the digital plane [4]. We note that the following implication holds for a point in a digital set:

$$\text{8-simple and perfect} \Rightarrow \text{4-simple.}$$

During this text we denote for illustration purposes the different types of points by the following symbols: '●' denotes a black point, 'o' a white point and '·' a point of either color.

3. Thinning

Rosenfeld [13] observed that the question whether a specific point P of a digital set S has influence on the "topology" of S or not, can be decided locally. More specifically, Rosenfeld proved the following Theorem:

Theorem 1. *The following properties of the point P of S are equivalent:*

- *P is a simple point of S*
- *$S \backslash \{P\}$ has the same number of components (in the S sense) as S, and $\mathbb{C}S \cup \{P\}$ has the same number of components (in the $\mathbb{C}S$ sense) as $\mathbb{C}S$.*

This Theorem is the theoretical justification for thinning methods. Such methods have in general the following structure:

Sequential Thinning Algorithm

    repeat
        Identify a simple point P in S **[Identification]**
        $S := S \backslash \{P\}$ **[Elimination]**
    until S contains no more simple points.

By means of this simple algorithm, a set is iteratively reduced to a "thin" set which is "topologically" equivalent to the original set. However, this algorithm does not lead to a well-defined method, since its result depends on the succession the simple points are identified. As a consequence, this method is of course not invariant with respect to motions mapping the digital plane onto itself. Specifically, the method cannot be implemented in parallel in this simple way since there may exist pairs of points being both simple, however, one of them remains no longer simple if the other is eliminated.

It is possible to formulate a well-defined invariant parallel method by making use of the concept of a perfect point. It is easily seen that the following method is topologically correct (i.e. all sets obtained in this way have the same number of 8-components and the complements of these sets have the same number of 4-components).

Invariant Thinning Algorithm

> **repeat**
>   **for** each point $P \in S$ **do in parallel**
>   **If** P is simple and perfect **then**
> P is labeled   **[Identification]**
>   **end if**
>   **end do**
>   **for** each labeled point P **do in parallel**
> $S := S \setminus \{P\}$ **[Elimination]**
>   **end do**
> **until** S contains no more simple and perfect points.

The method is formulated as a parallel method. This means that the Identification and the Elimination steps of the algorithm are performed for each point independently of the other points. Moreover, the method is invariant. By virtue of this latter property it is possible to prove a number of useful assertions concerning the properties of the method [4]. The set which is obtained from S by applying the Invariant Thinning Algorithm is uniquely determined. It is termed the *skeleton* of S and is denoted $\Sigma^I(S)$.

Based on the Invariant Thinning Method we formulate the following labeling method [5]:

Labeling Algorithm

> All points in $\complement S$ receive label 0
> All boundary points of S receive label 1   **[Initialization]**
> **repeat**
>   **for** each point P on the inner boundary **do in parallel**
>   Let l be the smallest label
>   among the labeled direct neighbors of P
> P receives label $l + 1$ **[Labeling]**
>   **end do**
>   **for** each point P which is simple and perfect **do**
>   Let l be the label of P
>   **if** there is a perfect partner of P
>   which has label $l + 1$ **then**
> $S := S \setminus \{P\}$ **[Elimination]**
>   **end do**
> **until** S contains no more points which can be eliminated.

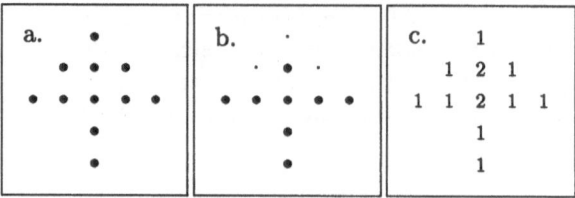

Figure 2. a Digital set containing two interior points. **b** First thinning step. In the second step one of the interior points of the original set will be deleted. **c** Labels of points according to the Labeling Algorithm. Both interior points obtain label 2, hence they are not eliminated

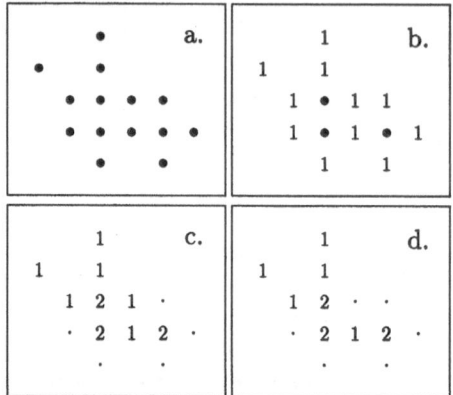

Figure 3. a Digital set. **b** Initialization. **c** First labeling and elimination step. **d** Second elimination step. The lower left point labeled 2 is not deleted by the Labeling Algorithm. This means that this point belongs to $\Sigma^l(S)$ but not to $\Sigma^l(S)$

This labeling method (with a slight modification) was described in [4] and [5]. The set which is obtained from S by applying the Labeling Algorithm is denoted $\Sigma^l(S)$. In general it is different from $\Sigma^l(S)$. In Fig. 3 an example is given for a simple and perfect point which is not eliminated by the labeling process. In Fig. 2 there is an interior point which belongs to $\Sigma^l(S)$ although it is a partner of a simple and perfect point.

If a point obtains label l by the labeling method, then there exists a 4-path of length l from this point to the complement of S. If the method proposed in [4, 5] is applied, this path is not necessarily a shortest path (see Fig. 4).

We can easily see that the original set can be reconstructed from the labeled skeleton. The reconstruction algorithm can be formulated as follows:

Reconstruction Algorithm

Let l be the highest possible label in $\Sigma^l(S)$
Let $S := \Sigma^l(S)$ **[Initialization]**

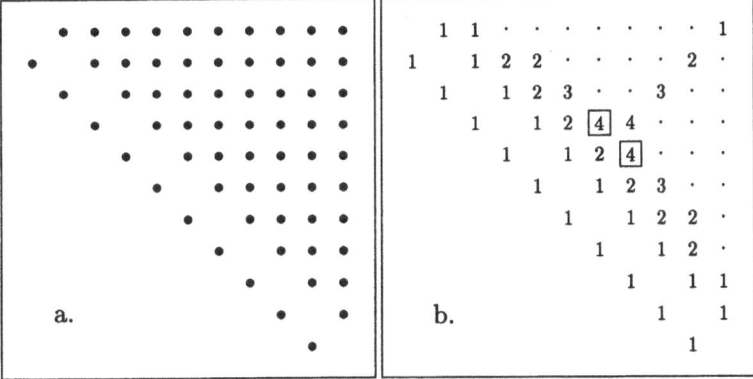

Figure 4. a Digital set. **b** Skeleton labeled by the method in [4, 5]. The two points labeled ④ have distances 3 to the complement of S

Repeat
    for all points $P \in S$ having label l **do in parallel**
        for all direct neighbors Q of P **do**
            if $Q \notin S$ **then** $S := S \cup \{Q\}$ **[Reconstruction]**
                Q obtains label $l - 1$ **[Update Label]**
            end if
        end do
    end do
    $l := l - 1$
until $l = 1$.

For further reference we define recursively the sequence of sets obtained as intermediate steps of the Labeling Algorithm. Let $S_0 := S$, and S_{l+1} is obtained by removing from S_l points according the the Labeling Algorithm. Then obviously $S_{l+1} \subseteq S_l$ and for finite S there is an l_0 such that $S_l = S_{l_0}$ for all $l \geq l_0$.

4. Maximal 4-Disks

For points $P = (i,j)^\top$ and $Q = (k,l)^\top$ in the digital plane we define the 4- or L_1-*distance* by

$$d_4(P,Q) = |i - k| + |j - l|.$$

The 4-*disk* or shortly *disk* with *center* $P \in \mathbb{Z}^2$ and *radius* $r \in \mathbb{N}$ is the set

$$B_r(P) = \{Q \in \mathbb{Z}^2 \,|\, d_4(P,Q) \leq r\}.$$

All points within a 4-disk are contained in a rhombus which has four quadrants, sides and vertices as indicated in Fig. 5. The sides are understood not to contain the adjacent vertices and similarly, the quadrants do not contain the sides. If we want to make clear that a quadrant, side or vertex belongs to $B_r(P)$ we write $q_i^{(r)}(P)$,

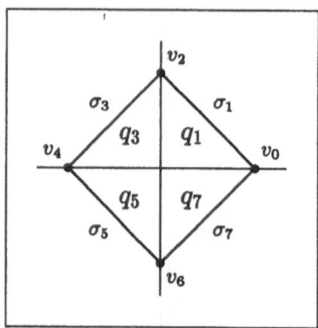

Figure 5. Continuous illustration of a digital rhombus (4-disk) with quadrants q_i and sides σ_i, $(i = 1, 3, 5, 7)$, and vertices v_i, $(i = 0, 2, 4, 6)$

$\sigma_i^{(r)}(P)$ or $v_i^{(r)}(P)$, respectively. Sometimes we need the *closed sides*

$$\bar{\sigma}_i^{(r)}(P) := v_{i-1}^{(r)}(P) \cup \sigma_i^{(r)}(P) \cup v_{i+1}^{(r)}(P).$$

In analogy to the indexing of neighborhood points we understand the lower indices mod 8.

Given a digital set $S \subseteq \mathbb{Z}^2$. We assume that S is bounded. A disk which is contained in the set S is termed *maximal* if there is no disk which is contained in S and which strictly contains the disk given. The set of all centers of maximal disks of a set S is denoted $M(S)$.

For the following Theorem it is useful to carry in mind that

$$B_{r+1}(P) = \bigcup_{i=0,2,4,6} B_r(N_i(P)).$$

Theorem 2. *Given a digital set S, $B_r(P) \subseteq S$, however $B_{r+1} \not\subseteq S$. $B_r(P)$ is not maximal if and only if there exist two adjacent sides of $B_{r+1}(P)$ and $B_{r+2}(P)$, namely $\sigma_{2k-1}^{(r+1)}(P)$, $\sigma_{2k+1}^{(r+1)}(P)$ and $\sigma_{2k-1}^{(r+2)}(P)$, $\sigma_{2k+1}^{(r+2)}(P)$, and corresponding vertices $v_{2k}^{(r+1)}(P)$, $v_{2k}^{(r+2)}(P)$ between them, which all are completely contained in S.*

Proof: 1. Assume that the conditions of the Theorem hold. To be specific, $\sigma_3^{(r+1)}(P)$, $\sigma_3^{(r+2)}(P)$, $\sigma_5^{(r+1)}(P)$, $\sigma_5^{(r+2)}(P)$, $v_4^{(r+1)}(P)$, $v_4^{(r+2)}(P)$ are all contained in S. Then it is easily seen that $B_r(P) \subseteq B_{r+1}(N_4(P)) \subseteq S$, hence $B_r(P)$ is not maximal.

2. Assume that $B_r(P)$ is not maximal. Each disk which strictly contains $B_r(P)$ also contains at least one of the disks $B_{r+1}(N_k(P))$, $k = 0, 2, 4, 6$. From this we are led to the conditions of the Theorem. □

It is more complicated to characterize maximal disks. For sake of completeness we include such a characterization. We start with some definitions. Given a digital set S, $B_r(P) \subseteq S$, however, $B_{r+1}(P) \cap CS \neq \emptyset$.

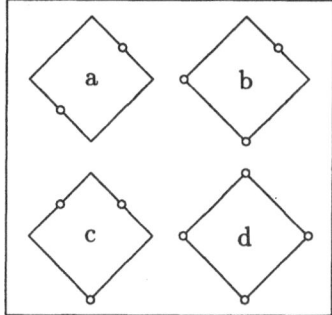

Figure 6. The four possible cases for a maximal rhombus. The elements of the rhombus which are supporting or blocking, respectively, are indicated by 'o'

A *supporting element* corresponding to P is either a *supporting side* $\sigma_i^{(r+1)}(P)$, $i = 1, 3, 5, 7$, which contains points from $\mathbb{C}S$ or a *supporting vertex* $v_i^{(r+1)}$, $i = 0, 2, 4, 6$, which is not in S.

A *k-blocking element*, $k = 0, 2, 4, 6$, corresponding to P is a supporting element of $N_k(P)$.

Theorem 3. $B_r(P)$ *is maximal if and only if one of the following conditions are fulfilled:*

1. *One side* $\sigma_{2i+1}^{(r+1)}(P)$ *is supporting and one of the following conditions holds:*

   (a) *The opposite side with index* $2i + 5$ *is supporting or* $2i + 4$- *or* $2i + 6$-*blocking (case a in Fig. 6).*

   (b) *The vertices with index* $2i + 4$ *and* $2i + 6$ *are supporting or* $2i + 4$- *or* $2i + 6$-*blocking (case b in Fig. 6).*

   (c) *One of the adjacent sides and the opposite vertex (i.e. side with index* $2i - 1$ *and vertex with index* $2i + 4$ *or side* $2i + 3$ *and vertex with index* $2i + 6$*) are supporting or* $2i + 4$- *or* $2i + 6$-*blocking (case c in Fig. 6).*

2. *One vertex* $v_{2i}^{(r+1)}(P)$ *is supporting and one of the following conditions holds:*

   (a) *The two opposite sides with index* $2i + 3$ *and* $2i + 5$ *are supporting or* $2i + 2$-, $2i + 4$- *or* $2i + 6$-*blocking (case c in Fig. 6).*

   (b) *The adjacent vertex and one opposite side (i.e. the vertex with index* $2i + 2$ *and the side with index* $2i + 5$ *or the vertex with index* $2i - 2$ *and the side with index* $2i - 5$*) are supporting or* $2i + 2$-, $2i + 4$- *or* $2i + 6$-*blocking (case b in Fig. 6).*

   (c) *All other vertices are supporting or* $2i + 2$-, $2i + 4$- *or* $2i + 6$-*blocking (case d in Fig. 6).*

For a point $P \in S$, the *(4-) distance of* P to the complement $\mathbb{C}S$ of S is the number

$$d(P, \mathbb{C}S) = \inf_{Q \in \mathbb{C}S} d_4(P, Q).$$

The set

$$\Pi_{\mathbb{C}S}(P) = \{Q \in \mathbb{C}S \,|\, d_4(P, Q) = d(P, \mathbb{C}S)\}$$

is termed the *metric* projection of P onto $\mathbb{C}S$.

Given a point $P \in S$ such that $d(P, \mathbb{C}S) = r + 1$. P is termed *regular*, if all sets $\sigma_i^{(r+1)}(P) \cap \mathbb{C}S$, $i = 1, 3, 5, 7$, are 8-connected. If a point in S is not regular, it is termed *singular*.

In (continuous) normed spaces with strictly convex and smooth norms a singular point can be defined as the center of a disk contained in S which meets the boundary of S in more than one point. It is well known that in this case each singular point is the center of a maximal disk and vice versa, the center of a maximal disk is contained in the topological closure of the set of singular points. These two assertions are known as Motzkin's Theorems [10, 11]. Moreover, in such spaces one can define the skeleton of a set to be the set of all centers of maximal disks. We saw from Fig. 4 that this is not necessarily true in our situation.

Whereas the centers of maximal disks or singular points can be easily determined from the *distance transformation* of S which is a function assigning to each point in S its distance from the complement, thinning is an iterative method which is very time consuming in two dimensions and needs a prohibitive amount of time in three dimensions. On the other hand, the distance map can be easily constructed by means of a two-phase sweep through S. This is the reason that Neusius and Olszewski [12] proposed a thinning method essentially using the distance map. It should be noted, however, that the above remarks apply only for the case of sequential thinning. For parallel thinning the picture is different.

It is obviously easily possible to reconstruct a digital set whenever one knows the centers of all maximal disks contained in it together with the corresponding radii. The reconstruction algorithm in this case becomes (in principle) extremely simple:

Reconstruction from Centers of Maximal Disks

> Let $S := M(S)$ [Initialization]
> **for** all centers P of maximal disks with radius r **do in parallel**
> $S := S \cup \mathcal{N}_4(P)$ [Reconstruction]
> **end do**

5. Thinning Well-Composed Regular Sets

The set $M(S)$ of all centers of maximal disks can be easily determined constructively by means of the distance transformation of S. Moreover, it exhibits in a natural way nice invariance properties. Neusius and Olszewski [12] therefore determined in a first step the so-called *preskeleton*, which is a subset of $M(S)$ (It is, however, not invariant, see [12, Figure 3]). On the other hand, the set $M(S)$ does

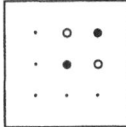

Figure 7. Critical configuration for non-well-composed sets

not have the same connectedness properties as the original set S (see Figure 8). Therefore it is necessary to "interpolate" the connected components of $M(S)$ in such a way that the resulting set is topologically equivalent to S.

In order to derive relations between Σ^l and $M(S)$, we need in addition to the regularity condition one more assumption about the set S.

In [8] the concept of a well-composed set was introduced and it was shown that for such sets certain peculiarities cannot happen. A digital set is termed a *well-composed set* if it does not contain the *critical configuration* of Fig. 7 (or the configuration obtained therefrom by a 90° rotation).

We note that in Fig. 4 as well as in Arcelli's sets in Fig. 1 critical configurations occur.

The aim of this section is to prove the following Theorem:

Theorem 4. *If S is a well-composed set and each point of S is a regular point then $M(S) \subseteq \Sigma^l(S)$.*

The proof is a consequence of the following Lemma.

Lemma 1. *Let S be a well-composed set, P a regular point in S, $B_r(P) \subseteq S$ and $\sigma_i^{(r+1)}(P) \cap CS \neq \emptyset$.*

Then the following assertions are true for $l = 1, 2, \ldots, r-1$:

(1) $B_{r-l}(P) \subseteq S_l$.
(2) $\sigma_i^{(r-l+1)}(P) \cap CS_l$ *is nonempty and 8-connected.*
(3) *P is a regular point in S_l and $S_l \cap [\sigma_i^{(r-l)} \cup \sigma_i^{(r-l+1)} \cup \sigma_i^{(r-l+2)}]$ is a well composed set.*

Proof: The set $\sigma_i^{(r+1)}(P)$ contains at most one 8-component of CS. Each pair of successive white points in $\sigma_i^{(r+1)}(P)$ has a common direct neighbor in CS by virtue of well-composedness. Thus, $S_0 \cap B_{r+2}(P)$ is well-composed.

Without loss of generality we will treat the side $\sigma_1^{(r+1)}(P)$.

Assume that the first three conditions hold for an $l < r-1$. We show that these conditions hold also for $l+1$.

It is clear that $B_{r-l-1}(P)$ is contained in S_{l+1}.

Assume that $S_{l+1} \cap [\sigma_1^{(r-l-1)} \cup \sigma_1^{(r-l)} \cup \sigma_1^{(r-l+1)}]$ is not a well-composed set. Then we have a black point Q in $\sigma_1^{(r-l-1)}(P)$ which has two direct white neighbors $N_0(Q)$ and $N_2(Q)$ in $\sigma_1^{(r-l)}(P)$ and these in turn have a common direct black neighbor $N_1(Q)$ in $\sigma_1^{(r-l+1)}(P)$. Since the white points in $\sigma_1^{(r-l)}(P)$ were created in the $(l+1)$rst step, there exist two white components in $\sigma_1^{(r-l+1)}(P)$ which are 4-connected to these points. These components are separated by $N_1(Q) \in \sigma_1^{(r-l+1)}(P)$ which contradicts regularity.

By well-composedness (as assumed in (3)) each perfect point on $\sigma_1^{(r-l+1)}(P)$ is simple. Each pair of successive white points on $\sigma_1^{(r-l+1)}(P)$ has a common direct neighbor which is perfect. All these neighbors are on $\sigma_1^{(r-l)}(P)$ and constitute an 8-connected set. If Q is a white point on $\sigma_1^{(r-l+1)}(P)$ which has only one white neighbor on $\sigma_1^{(r-l+1)}(P)$ then Q shares a perfect point with its white neighbor and possibly has another direct neighbor which is perfect. Certainly, in the latter case both perfect neighbors are 8-neighbors of each other. If there is only one white point on $\sigma_1^{(r-l+1)}(P)$ then it has one or two perfect direct neighbors.

The possible situations are illustrated in the following picture. White points on $\sigma_1^{(r-l+1)}(P)$ are joined by arrows with those they make perfect on $\sigma_1^{(r-l)}(P)$.

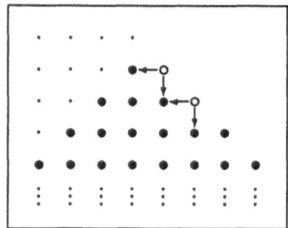

We conclude: All perfect points are simple by well-composedness of S. They all are on $\sigma_1^{(r-l)}(P)$ and form an 8-connected set. To each pair of adjacent perfect points there is a common direct neighbor on $\sigma_1^{(r-l+1)}(P)$ in $\mathbb{C}S$. This proves (2) and in particular regularity. $\qquad\square$

Remark. It can happen that an S_l contains a critical configuration. However, such a configuration can only occur in such a way that one of the white points of it coincides with a vertex $v_i^{(r-l+1)}(P)$ of $B_{r-l+1}(P)$. In the following example the critical configuration in the skeleton (left) occurs in the second thinning step and involves the vertex $v_2^{(2)}(P)$.

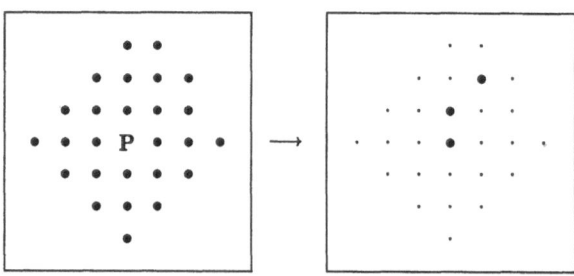

As an immediate consequence of the Lemma we state:

Theorem 5. *Let S be a well-composed set, $P \in S$, $B_r(P) \subseteq S$ and $B_{r+1}(P) \cap CS \neq \emptyset$. Assume further that P and its four direct neighbors are regular points in S.*

$$P \in M(S) \Rightarrow P \in \Sigma^l(S).$$

Proof: For $P \in M(S)$ then we investigate all possibilities of Theorem 3. We only consider here one typical case for illustration purposes.

Assume that only $\sigma_1^{(r+1)}(P)$ and $\sigma_1^{(r+1)}(N_4(P))$ contain points in CS (case a in Fig. 6). Then, according to the Lemma, after $r - 1$ thinning steps $\sigma_1^{(2)}(P) \cap CS_{r-1} \neq \emptyset$ and $\sigma_5^{(2)}(N_4(P)) \cap CS_{r-1} \neq \emptyset$. This means that $N_1(P)$ and $N_5(N_4(P))$ are eliminated and they obtain label $r - 1$.

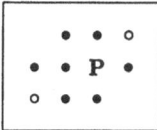

Again by the Lemma, the white points $N_1(P)$ and $N_5(N_4(P))$ both carry the same label $r - 1$. Hence, P and $N_4(P)$ have label $r + 1$. Similarly, all direct neighbors of P receive a label r or $r + 1$. This implies that P is not eliminated by the Labeling Algorithm. □

Remark. An alternative approach to thinning is to determine first $M(S)$ and to apply then any thinning method with the additional requirement that points in $M(S)$ are never eliminated. Such an approach was proposed in [6]. The reason for this proposal was to retain the favourable properties of $M(S)$ and simultaneously to have 'correct topology'. While in this method the assertion of the Theorem was forced, we see here that we get it without any additional effort.

The Theorem states that the skeleton $\Sigma^l(S)$ which is obtained by the Labeling Algorithm "interpolates" the set $M(S)$ as to yield a set having the same connectedness properties as the original set S.

Under the assumption that each point in S is regular we can state $M(S) \subseteq \Sigma^l(S)$. This proves Theorem 4. Without the condition of regularity this assertion is not generally true. It can happen that points in $M(S)$ do not necessarily belong to $\Sigma^l(S)$. In Fig. 8 examples of such points are presented.

Of course, if S is not well-composed, the assertion of Theorem 4 is also not generally true as demonstrated in Fig. 4.

The question remains open whether it is possible to characterize those points in $\Sigma^l(S)$ which are not in $M(S)$.

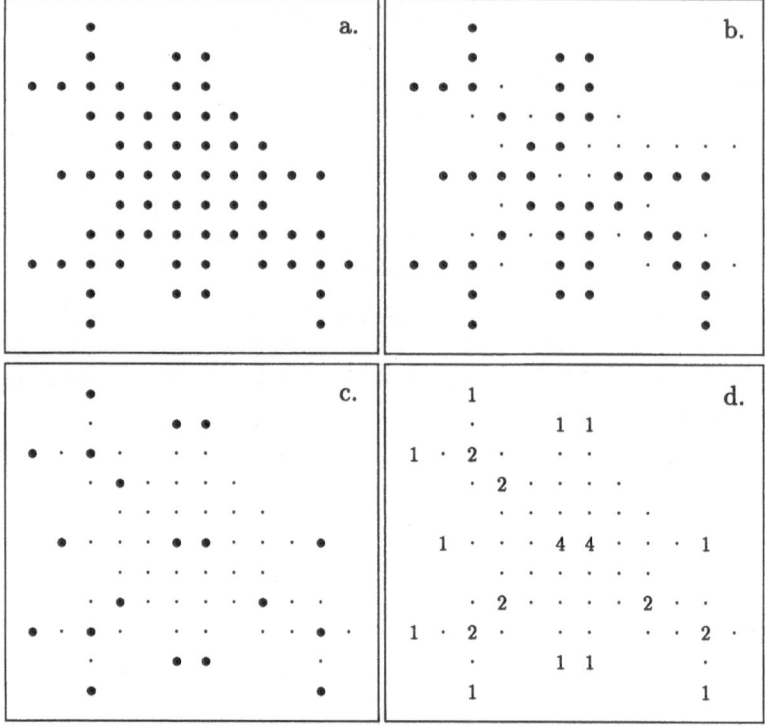

Figure 8. a Digital set. **b** Skeleton. **c** Centers of maximal 4-disks. **d** Distances of centers of maximal disks from the complement

6. Conclusions

The theory presented in this paper covers only the regular case. In this case, a very simple and practically easily verifiable local condition, the condition of well-composedness, guarantees that $M(S) \subseteq \Sigma^1(S)$.

However, regularity is a global condition which cannot be checked by a simple algorithm. As the example in Fig. 8 demonstrates, the regular case is not representative. It is possible to treat also the singular case, however, much more technical preparations are needed which are beyond the scope of this paper.

One might ask whether practical conclusions for the implementation of thinning algorithms can be drawn from the results given here. Thinning can be characterized as a widely used preprocessing method which has a remarkable lack of theory. Specifically, there is almost no bridge from the theory of the continuous medial axis transform to the skeletons obtained by thinning. Neusius and Olszewski [12] showed that the discrete analog of the medial axis transform which in general has not the same 'topology' as the original set can be made to a 'deformation retract' of the latter. In this paper we showed that in the regular case a certain variant of the skeleton of a set contains the medial axis. So it might be possible to relate the discrete theory to the rich and developed continuous theory.

In a paper of Bardtke and Berens [2] it was shown that the skeleton of a plane set can be obtained by solving a differential equation. Independently from this idea different authors [17, 6] proposed to construct the skeleton in an incremental way: Given a skeleton point, they tried to construct neighboring skeleton points by simultaneously identifying large regions which do not belong to the skeleton. These approaches were indeed very efficient, however, due to the lack of a sound theoretical background, the results were not always convincing. It would be interesting to reinvestigate such approaches.

The thinning method which was used here is not new, there exists a widely distributed implementation and large amounts of numerical and practical experiences. Details can be found in [4].

References

[1] Arcelli, C.: Pattern thinning by contour tracing. Comput. Graphics Image Proc. *17*, 130–144 (1981).

[2] Bardtke, K., Berens, H.: Eine Beschreibung der Nichteindeutigkeitsmenge für die beste Approximation in der Euklidischen Ebene. J. Approx. Theory *47*, 54–74 (1986).

[3] Eckhardt, U., Latecki, L., Maderlechner, G.: Irreducible and thin binary sets. 2nd International Workshop on Visual Form, Capri, Italy, May 30–June 2, 1994.

[4] Eckhardt, U., Maderlechner, G.: Invariant thinning. Int. J. Pattern Rec. Art. Intell. *7*, 1115–1144 (1993).

[5] Eckhardt, U., Maderlechner, G.: Thinning binary pictures by a labeling procedure. Proceedings 11th IAPR International Conference on Pattern Recognition, The Hague, The Netherlands, August 30–September 3, 1992, Volume III, Conference C: Image, Speech, and Signal Analysis, pp. 582–585. Los Alamitos, Washington, Brussels, Tokyo: IEEE Computer Society Press 1992.

[6] Evers, C., Andersen, J., Maderlechner, G.: Ein neues Verfahren zur euklidischen Skelettierung von Binärbildern. In Mustererkennung 1987. 9. DAGM-Symposium, Braunschweig, 29.9.–1.10. 1987, Proceedings, Informatik-Fachberichte 149 (Paulus, E., ed.), p. 171. Berlin, Heidelberg, New York, Tokyo: Springer. 1987.

[7] Haralick, R. M.: Performance characterization in image analysis: thinning, a case in point. Pattern Rec. Lett. *13*, 5–12 (1992).

[8] Latecki, L., Eckhardt, U., Rosenfeld, A.: Well-Composed Sets. Comput. Vision Image Underst. *61*, 70–83 (1995).

[9] Minsky, M., Papert, S.: Perceptrons. An introduction to computational geometry. Cambridge Mass., London: The MIT Press 1969.

[10] Motzkin, T.: Sur quelques propriétés caractéristiques des ensembles bornés non convexes. Atti della Reale Accademia Nazionale dei Lincei, Serie sesta, Rendiconti, Classe di Scienze Fisiche, Matematiche e Naturali *21*, 773–779 (1935).

[11] Motzkin, T.: Sur quelques propriétés caractéristiques des ensembles convexes. Atti della Reale Accademia Nazionale dei Lincei, Serie sesta, Rendiconti, Classe di Scienze Fisiche, Matematiche e Naturali *21*, 562–567 (1935).

[12] Neusius, C., Olszewski, J.: A noniterative thinning algorithm. ACM Trans. Math. Software *20*, 5–20 (1994).

[13] Rosenfeld, A.: Digital topology. Amer. Math. *86*, 621–630 (1979).

[14] Serra, J.: Image analysis and mathematical morphology. London, New York: Academic Press 1982.

[15] Serra, J., ed.: Image analysis and mathematical morphology, Vol. 2: Theoretical advances. London, New York: Academic Press 1988.

[16] Yu, Z., Conrad, C., Eckhardt, U.: Regularization of the medial axis transform. In: Theoretical Foundations of Computer Vision 1992, Proceedings of the Vth Workshop 1992, Buckow (Märkische Schweiz), March 30–April 3, 1992, (Mathematical Research, Volume 69) (Klette, R., Kropatsch, W. G., eds.), pp. 13–24. Berlin: Akademie-Verlag 1992.
[17] Yu, Z.: Intrinsic topology of medial axis. In: Mustererkennung 1989. 11. DAGM-Symposium, Hamburg, 2.–4. Oktober 1989. Proceedings. Informatik-Fachberichte 219 (Burkhardt, H., Höhne, K. H., Neumann, B., eds.), pp. 77–81. Berlin, Heidelberg, New York, Tokyo: Springer 1989.

U. Eckhardt
Department of Applied Mathematics
University of Hamburg
Bundesstrasse 55
D-20146 Hamburg
Germany
e-mail: eckhardt@math.uni-hamburg.de

L. Latecki
Department of Computer Science
University of Hamburg
Vogt-Kölln-Strasse 30
D-22527 Hamburg
Germany
e-mail: latecki@informatik.uni-hamburg.de

Computing Suppl. 11, 37–51 (1996)

Recognition of Images Degraded by Linear Motion Blur without Restoration*

J. Flusser, T. Suk, and S. Saic, Prague

Abstract

Recognition of Images Degraded by Linear Motion Blur without Restoration. The paper is devoted to the feature-based description of images degraded by linear motion blur. The proposed features are invariant with respect to motion velocity, are based on image moments and are calculated directly from the blurred image. In that way, we are able to describe the original image without the PSF identification and image restoration. In many applications (such as in image recognition against a database) our approach is much more effective than the traditional "blind-restoration" one. The derivation of the motion blur invariants is a major theoretical result of the paper. Numerical experiments are presented to illustrate the utilization of the invariants for blurred image description. Stability of the invariants with respect to additive random noise is also discussed and is shown to be sufficiently high. Finally, another set of features which are invariant not only to motion velocity but also to motion direction is introduced.

Key words: Blurred image, linear imaging system, motion blur, image moments, blur invariants.

1. Introduction

The images of a moving scene or the images of a static scene taken by moving sensor represent the original objects mostly in an unsatisfactory manner. Due to the relative motion of the sensor and the scene and finite exposure time, such images are degraded by well-known degradation factor called *motion blur*.

Widely accepted standard linear model [1] describes the imaging process by convolution of an unknown original (or ideal) image $f(x, y)$ with space-invariant point spread function (PSF) $h(x, y)$:

$$g(x, y) = (f * h)(x, y) \tag{1}$$

where $g(x, y)$ represents the observed image. The PSF $h(x, y)$ describes the imaging system and it is supposed to be unknown.

In case of linear horizontal motion, $h(x, y)$ has the following form:

---

* This work has been supported by grant No. 102/94/1835 of the Grant Agency of the Czech Republic.

$$h(x, y) = \begin{cases} \dfrac{1}{vt} \delta(y) \Leftrightarrow 0 \leq x \leq vt \\[2mm] 0 \qquad \text{otherwise} \end{cases} \tag{2}$$

(v is the motion velocity, t is the exposure and δ is Dirac function).

Classical "blind-restoration" approach to dealing with motion blurred images consists of two following steps:

- Estimation of the PSF $h(x, y)$, i.e. estimation of the constant $c = vt$;
- Estimation of the ideal image $f(x, y)$ via restoration of the blurred image $g(x, y)$.

Both these steps have been dealt with extensively in literature during last two decades.

One group of methods for PSF identification is based on the investigation of zero patterns in the frequency domain [2]–[4] or spike patterns in the cepstral domain [5]. It is well-known [11] that there is an infinite number of parallel lines of zeros in Fourier spectrum of a motion blurred image, which are perpendicular to the motion direction. Knowing the distance between those lines, we can estimate the motion velocity.

Another group of methods is based on modeling of the image by a stochastic process. The original image is modeled as an AR process and the blur as a MA process. The blurred image is then modeled as an ARMA process and the MA process identified by this model is considered as a description of the PSF. In this way the problem of PSF estimation is transformed onto the problem of determining the parameters of an ARMA model [6]–[10].

After the PSF has been identified, the original image can be estimated via restoration of the blurred image by inverse filter, Wiener filter or by any other similar technique (see [1] and [11] for a survey).

Generally speaking, the above-mentioned approach to image restoration is very complicated and time-consuming. In many cases, we do not need to know the original image itself, but we have only to know some representation of it (a typical example is a recognition of a blurred image against a database.) However, such a representation should be independent of the motion velocity and direction and should describe really the original image, not the degraded one. In this paper, we introduce a set of features for image description which are invariant to motion blur (that means the feature values of $g(x, y)$ do not depend on the motion characteristics and they are the same as the feature values of $f(x, y)$). Image recognition may be then accomplished via classification in the feature space. In this way, we get rid of the necessity of the PSF identification and image restoration.

Motion blur invariant features introduced in this paper are based on image moments. In Section 2 we deal with ordinary and central moments of a blurred image and we express them as functions of moments of an ideal image and the PSF. Sections 3 and 4 perform the major part of the paper. An original algorithm for invariants derivation is presented and the invariants up to the 7th order are shown in explicit form. To demonstrate the performance of the invariants, numerical experiment is presented in Section 5. Section 6 is devoted to the invariant features in case of non-horizontal motion. Basic properties of the invariants (stability, uniqueness and efficiency) are briefly discussed in Section 7.

2. Blurred Image Moments

The two-dimensional $(p + q)$th order *ordinary moment* $m_{pq}^{(f)}$ of image $f(x, y)$ is defined by the integral

$$m_{pq}^{(f)} = \int_{-\infty}^{\infty} \int_{-\infty}^{\infty} x^p y^q f(x, y) \, dx \, dy. \tag{3}$$

Point $(x_t^{(f)}, y_t^{(f)})$ given by the relations

$$x_t^{(f)} = \frac{m_{10}^{(f)}}{m_{00}^{(f)}},$$

$$y_t^{(f)} = \frac{m_{01}^{(f)}}{m_{00}^{(f)}}$$

is called *center of gravity* or *centroid* of image $f(x, y)$. The $(p + q)$th order *central moment* $\mu_{pq}^{(f)}$ is then defined as

$$\mu_{pq}^{(f)} = \int_{-\infty}^{\infty} \int_{-\infty}^{\infty} (x - x_t^{(f)})^p (y - y_t^{(f)})^q f(x, y) \, dx \, dy. \tag{4}$$

The following two theorems describe how to express moments of the blurred image in terms of moments of the original image and the PSF.

Theorem 1. *Let $f(x, y)$ and $h(x, y)$ be piecewise continuous functions which are non-zero only on bounded supports. Let $g(x, y)$ be given by the convolution*

$$g(x, y) = (f * h)(x, y).$$

Then the relation

$$m_{pq}^{(g)} = \sum_{k=0}^{p} \sum_{j=0}^{q} \binom{p}{k} \binom{q}{j} m_{kj}^{(f)} m_{p-k, q-j}^{(h)}$$

holds for every p and q.

Proof:

$$m_{pq}^{(g)} = \int_{-\infty}^{\infty} \int_{-\infty}^{\infty} x^p y^q g(x, y) \, dx \, dy = \int_{-\infty}^{\infty} \int_{-\infty}^{\infty} x^p y^q (f * h)(x, y) \, dx \, dy$$

$$= \int_{-\infty}^{\infty} \int_{-\infty}^{\infty} x^p y^q \left(\int_{-\infty}^{\infty} \int_{-\infty}^{\infty} h(a, b) f(x - a, y - b) \, da \, db \right) dx \, dy$$

$$= \int_{-\infty}^{\infty} \int_{-\infty}^{\infty} h(a,b) \left(\int_{-\infty}^{\infty} \int_{-\infty}^{\infty} x^p y^q f(x-a, y-b) \, dx \, dy \right) da \, db$$

$$= \int_{-\infty}^{\infty} \int_{-\infty}^{\infty} h(a,b) \left(\int_{-\infty}^{\infty} \int_{-\infty}^{\infty} (x+a)^p (y+b)^q f(x,y) \, dx \, dy \right) da \, db$$

$$= \int_{-\infty}^{\infty} \int_{-\infty}^{\infty} h(a,b) \left(\sum_{k=0}^{p} \sum_{j=0}^{q} \binom{p}{k} \binom{q}{j} a^{p-k} b^{q-j} m_{kj}^{(f)} \right) da \, db$$

$$= \sum_{k=0}^{p} \sum_{j=0}^{q} \binom{p}{k} \binom{q}{j} m_{kj}^{(f)} m_{p-k,q-j}^{(h)} \qquad \square$$

Lemma 1. *Let* $f(x,y)$ $g(x,y)$ *and* $h(x,y)$ *be the functions from Theorem 1. Then the centroid of* $g(x,y)$ *is given by the sum of centroids of* $f(x,y)$ *and* $h(x,y)$:

$$(x_t^{(g)}, y_t^{(g)}) = (x_t^{(f)} + x_t^{(h)}, y_t^{(f)} + y_t^{(h)}).$$

Proof:

$$x_t^{(g)} = \frac{m_{10}^{(g)}}{m_{00}^{(g)}} = \frac{m_{10}^{(f)} m_{00}^{(h)} + m_{00}^{(f)} m_{10}^{(h)}}{m_{00}^{(f)} m_{00}^{(h)}} = \frac{m_{10}^{(f)}}{m_{00}^{(f)}} + \frac{m_{10}^{(h)}}{m_{00}^{(h)}} = x_t^{(f)} + x_t^{(h)} \qquad \square$$

The proof for $y_t^{(g)}$ is similar.

Theorem 2. *Let all assumptions of Theorem 1 be valid. Then the relation*

$$\mu_{pq}^{(g)} = \sum_{k=0}^{p} \sum_{j=0}^{q} \binom{p}{k} \binom{q}{j} \mu_{kj}^{(f)} \mu_{p-k,q-j}^{(h)}$$

holds for every p *and* q.

Proof:

$$\mu_{pq}^{(g)} = \int_{-\infty}^{\infty} \int_{-\infty}^{\infty} (x - x_t^{(g)})^p (y - y_t^{(g)})^q g(x,y) \, dx \, dy$$

$$= \int_{-\infty}^{\infty} \int_{-\infty}^{\infty} (x - x_t^{(f)} - x_t^{(h)})^p (y - y_t^{(f)} - y_t^{(h)})^q (f * h)(x,y) \, dx \, dy$$

$$= \int_{-\infty}^{\infty} \int_{-\infty}^{\infty} (x - x_t^{(f)} - x_t^{(h)})^p (y - y_t^{(f)} - y_t^{(h)})^q$$

$$\times \left(\int_{-\infty}^{\infty} \int_{-\infty}^{\infty} h(a,b) f(x-a, y-b) \, da \, db \right) dx \, dy$$

$$= \int_{-\infty}^{\infty} \int_{-\infty}^{\infty} h(a,b)$$

$$\times \left(\int_{-\infty}^{\infty} \int_{-\infty}^{\infty} (x - x_t^{(f)} - x_t^{(h)})^p (y - y_t^{(f)} - y_t^{(h)})^q f(x-a, y-b) \, dx \, dy \right) da \, db$$

$$= \int_{-\infty}^{\infty} \int_{-\infty}^{\infty} h(a,b) \Bigg(\int_{-\infty}^{\infty} \int_{-\infty}^{\infty} (x - x_t^{(f)} + a - x_t^{(h)})^p (y - y_t^{(f)} + b - y_t^{(h)})^q$$

$$\cdot f(x,y)\, dx\, dy \Bigg) da\, db$$

$$= \int_{-\infty}^{\infty} \int_{-\infty}^{\infty} h(a,b) \left(\sum_{k=0}^{p} \sum_{j=0}^{q} \binom{p}{k}\binom{q}{j} (a - x_t^{(h)})^{p-k} (b - y_t^{(h)})^{q-j} \mu_{kj}^{(f)} \right) da\, db$$

$$= \sum_{k=0}^{p} \sum_{j=0}^{q} \binom{p}{k}\binom{q}{j} \mu_{kj}^{(f)} \mu_{p-k,q-j}^{(h)} \qquad \square$$

Lemma 2. *Let $h(x,y)$ be given by relation (2). Then*

- $\mu_{pq}^{(h)} = 0$ *for every* $q \neq 0$;
- *if p is odd, then $\mu_{p0}^{(h)} = 0$.*

The proof of Lemma 2 is straightforward.

Using Lemma 2 and Theorem 2, we get the following.

Lemma 3. *Let*

$$g(x,y) = (f * h)(x,y)$$

and let $h(x,y)$ be given by relation (2). Then

$$\mu_{pq}^{(g)} = \sum_{k=0}^{[p/2]} \binom{p}{2k} \mu_{p-2k,q}^{(f)} \mu_{2k,0}^{(h)}.$$

Lemma 4. *Let $h(x,y)$ be given by relation (2). Then*

$$\int_{-\infty}^{\infty} \int_{-\infty}^{\infty} h(x,y)\, dx\, dy = \mu_{00}^{(h)} = 1.$$

That means motion blur preserves global energy of the image.

3. Derivation of the Motion Blur Invariants

Motion blur invariants are supposed to be functions of image central moments, i.e. they are supposed to have a form $M = M(\mu_{00}, \mu_{20}, \ldots, \mu_{pq})$. By the order r of invariant M we understand the order of the highest moment μ_{pq}, i.e. $r = p + q$.

Derivation of the invariants up to the 3rd order is almost trivial. It is quite easy to prove by means of Lemma 3 that $\mu_{00}, \mu_{11}, \mu_{02}, \mu_{12}, \mu_{21}, \mu_{03}$ and μ_{30} are invariant with respect to linear motion blur.

We propose the following algorithm for the construction of motion blur invariants of order $r > 3$.

1. Let $r > 3$ be the order of desired invariant M. Let μ_{pq} be any central moment of order r (except μ_{r0} if r is even). Then we start with settings

$$K = [p/2] - 1$$

(symbol $[x]$ denotes integer part of x) and

$$I_0 = \mu_{pq}.$$

2. **for** $n = 0$ **to** K
   Define D_n as

$$D_n = I_n^{(g)} - I_n^{(f)}.$$

   D_n has the form

$$D_n = F_n(\mu^{(f)})\mu_{a_n 0}^{(h)} + R_n(\mu^{(f)}, \mu^{(h)})$$

   where F_n is a function of central moments of image $f(x, y)$ only and $\mu_{a_n 0}^{(h)}$ is a central moment of the highest order of $h(x, y)$. No moment of $h(x, y)$ of the same order is contained in $R_n(\mu^{(f)}, \mu^{(h)})$. Moments of $g(x, y)$ were evaluated by means of Lemma 3. It holds

$$a_n = 2(K - n + 1).$$

   Due to the properties of $h(x, y)$, a_n is even for every n. Define I_{n+1} as

$$I_{n+1} = I_n - \frac{1}{\mu_{00}} F_n(\mu)\mu_{a_n 0}.$$

   end for

3. Set

$$M = I_{K+1}.$$

Note that this is an algorithm for *construction* of invariants, not for their evaluation. The invariants are evaluated by means of their explicit forms given in the next Section.

However, it is not clear if every function M produced by this algorithm is really motion blur invariant. The answer is given by the following Lemma.

Lemma 5. *Let us consider the situation in Step 2 for the last n (i.e. $n = K$). Then*

- D_K *has the form*

$$D_K = F_K(\mu^{(f)})\mu_{20}^{(h)}.$$

- *If the function $F_K(\mu)$ is a motion blur invariant, then the function M produced as a result of the above described algorithm is also motion blur invariant.*

Proof:

- Proof of the first assertion is straightforward.

$$\bullet \qquad M^{(g)} = I^{(g)}_{K+1} = I^{(g)}_K - \frac{F_K(\mu^{(g)})\mu^{(g)}_{20}}{\mu^{(g)}_{00}}$$

$$= I^{(f)}_K + F_K(\mu^{(f)})\mu^{(h)}_{20} - \frac{F_K(\mu^{(g)})(\mu^{(f)}_{00}\mu^{(h)}_{20} + \mu^{(f)}_{20})}{\mu^{(g)}_{00}}$$

$$= I^{(f)}_K - \frac{F_K(\mu^{(f)})\mu^{(f)}_{20}}{\mu^{(f)}_{00}} = I^{(f)}_{K+1} = M^{(f)} \qquad \square$$

As you can see from the algorithm description, the number n_r of independent invariants of order r $(r > 1)$ is

$$n_r = r + 1$$

if r is odd (each moment μ_{pq} of order r generates one independent invariant) and

$$n_r = r$$

if r is even (moment μ_{r0} does not generate any invariant).

Since the invariants should serve as features for image similarity or dissimilarity assessment, it is sometimes inconvenient to use directly the invariants described above.

The invariants should be normalized in two ways: to be independent of average gray-level value of the image and to have the same "weight" in Euclidean metric space. To achieve this, we use *normalized motion blur invariants*

$$M' = \frac{M}{\mu_{00} \cdot (N/2)^r}$$

where N is the size of the image and r is the order of M.

The set $\mathcal{M}_r$ of invariants M' up to the rth order has

$$N_r = \sum_{k=2}^{r} n_k$$

elements. The N_r-dimensional metric space $(\mathcal{M}_r, \varrho)$ where ϱ is Euclidean metric is a suitable feature space for evaluating image-to-image distances.

4. Motion Blur Invariants in Explicit Form

Applying the above-described algorithm, we can construct motion blur invariants of any order and express them in explicit form. A set of the invariants up to the 7th order is listed below:

• Zero-order:

$$M_0 = \mu_{00}.$$

- 2nd order:

  $$M_1 = \mu_{11},$$

  $$M_2 = \mu_{02},$$

- 3rd order:

  $$M_3 = \mu_{12},$$

  $$M_4 = \mu_{21},$$

  $$M_5 = \mu_{03},$$

  $$M_6 = \mu_{30}.$$

- 4th order:

  $$M_7 = \mu_{04},$$

  $$M_8 = \mu_{13},$$

  $$M_9 = \mu_{22} - \frac{\mu_{20}\mu_{02}}{\mu_{00}},$$

  $$M_{10} = \mu_{31} - \frac{3\mu_{20}\mu_{11}}{\mu_{00}}.$$

- 5th order:

  $$M_{11} = \mu_{14},$$

  $$M_{12} = \mu_{05},$$

  $$M_{13} = \mu_{32} - \frac{3\mu_{20}\mu_{12}}{\mu_{00}},$$

  $$M_{14} = \mu_{23} - \frac{\mu_{20}\mu_{03}}{\mu_{00}},$$

  $$M_{15} = \mu_{41} - \frac{6\mu_{20}\mu_{21}}{\mu_{00}},$$

  $$M_{16} = \mu_{50} - \frac{10\mu_{20}\mu_{30}}{\mu_{00}}.$$

- 6th order:

  $$M_{17} = \mu_{06},$$

  $$M_{18} = \mu_{15},$$

  $$M_{19} = \mu_{24} - \frac{\mu_{20}\mu_{04}}{\mu_{00}},$$

  $$M_{20} = \mu_{33} - \frac{3\mu_{20}\mu_{13}}{\mu_{00}},$$

$$M_{21} = \mu_{42} - \frac{\mu_{40}\mu_{02} + 6\mu_{20}M_9}{\mu_{00}},$$

$$M_{22} = \mu_{51} - \frac{5(\mu_{40}\mu_{11} + 2\mu_{20}M_{10})}{\mu_{00}}.$$

- 7th order:

$$M_{23} = \mu_{07},$$

$$M_{24} = \mu_{16},$$

$$M_{25} = \mu_{25} - \frac{\mu_{20}\mu_{05}}{\mu_{00}},$$

$$M_{26} = \mu_{34} - \frac{3\mu_{20}\mu_{14}}{\mu_{00}},$$

$$M_{27} = \mu_{43} - \frac{\mu_{40}\mu_{03} + 6\mu_{20}M_{14}}{\mu_{00}},$$

$$M_{28} = \mu_{52} - \frac{5(\mu_{40}\mu_{12} + 2\mu_{20}M_{13})}{\mu_{00}},$$

$$M_{29} = \mu_{61} - \frac{15(\mu_{40}\mu_{21} + \mu_{20}M_{15})}{\mu_{00}},$$

$$M_{30} = \mu_{70} - \frac{35\mu_{40}\mu_{30} + 21\mu_{20}M_{16}}{\mu_{00}}.$$

5. Numerical Experiment

In order to demonstrate the performance of the proposed blurred image features, a number of experiments were carried out. Major goal of the experiment described below was to prove the invariance of the features with respect to motion blur and to test their stability under additive random noise.

The test images used for the experiment are shown in Fig. 1. From top to bottom and from left to right you can see original Lena image of the size 256×256 pixels, Lena image blurred by slow horizontal motion ($v \cdot t = 30$ pixels), the blurred image corrupted by additive zero-mean Gaussian noise with standard deviation 10 and 30, respectively and Lena image blurred by fast horizontal motion ($v \cdot t = 80$ pixels). Last image—portrait of Lisa—was incorporated into the test set to show the discriminability of the invariants.

The invariants $M_0, M_1, \ldots, M_{22}$ were calculated for each image. However, in case of digital image f_{ij} of the size $N \times N$ we have to use a discrete version of Eq. (4) for moment evaluation:

Figure 1. Top row: Lena original (left), horizontal motion blur ($v \cdot t = 30$ pixels) (middle), image b corrupted by additive Gaussian random noise (STD = 10) (right); bottom row: image b corrupted by additive Gaussian random noise (STD = 30) (left), horizontal motion blur ($v \cdot t = 80$ pixels) (middle), Lisa original (right)

$$\mu_{pq} = \sum_{i=1}^{N} \sum_{j=1}^{N} (i - x_t)^p (j - y_t)^q f_{ij}. \tag{5}$$

The values of invariants are summarized in Table 1. It is clearly visible from Table 1 that $M_0, M_1, \ldots, M_{22}$ are invariant to motion velocity, sufficiently stable under additive noise and object discriminatory.

6. Motion Blur Invariants of Rotated and Scaled Images

The invariants introduced in Section 3 are invariant to motion velocity and to image translation but their values depend on image orientation, scale and motion direction. This fact may cause problems in some applications.

To obtain invariants to image scaling

$$x' = s \cdot x,$$

$$y' = s \cdot y$$

it is sufficient only to replace central moments μ_{pq} by *normalized central moments* v_{pq} in definitions of invariants M_i. Normalized central moments are defined as

Table 1. The values of the invariants of the images in Fig. 1

	Lena orig.	Motion 30 pix.	Motion 30 STD = 10	Motion 30 STD = 30	Motion 80 pix.	Lisa
$M_0[10^6]$	6.4586	6.4589	6.4607	6.4674	6.4590	5.0860
$M_1[10^{10}]$	0.2560	0.2559	0.2553	0.2612	0.2560	−0.6036
$M_2[10^{10}]$	3.4740	3.4742	3.4750	3.4816	3.4743	2.3215
$M_3[10^{11}]$	−1.1022	−1.1024	−1.0968	−1.0616	−1.1017	1.4777
$M_4[10^{11}]$	1.0456	1.0462	1.0570	1.0639	1.0450	−2.9602
$M_5[10^{11}]$	2.2436	2.2434	2.2313	2.2541	2.2444	0.4081
$M_6[10^{11}]$	−5.3930	−5.3925	−5.3901	−5.4370	−5.3934	6.8344
$M_7[10^{14}]$	3.3918	3.3920	3.3918	3.4093	3.3921	2.0830
$M_8[10^{14}]$	0.1776	0.1775	0.1772	0.1811	0.1776	−0.5208
$M_9[10^{14}]$	−0.0513	−0.0513	−0.0507	−0.0544	−0.0513	0.0426
$M_{10}[10^{14}]$	−0.2293	−0.2292	−0.2293	−0.2277	−0.2292	0.5762
$M_{11}[10^{15}]$	−1.5087	−1.5088	−1.4996	−1.4809	−1.5081	2.7966
$M_{12}[10^{15}]$	4.7607	4.7606	4.7382	4.7451	4.7622	0.7743
$M_{13}[10^{15}]$	−3.1958	−3.1954	−3.1843	−3.2365	−3.1957	0.6806
$M_{14}[10^{15}]$	0.0734	0.0739	0.0985	0.0330	0.0732	−2.8659
$M_{15}[10^{16}]$	−0.2281	−0.2283	−0.2280	−0.2314	−0.2282	0.9275
$M_{16}[10^{16}]$	1.7525	1.7524	1.7540	1.7645	1.7528	−3.5346
$M_{17}[10^{18}]$	3.9858	3.9860	3.9817	4.0315	3.9863	2.3215
$M_{18}[10^{18}]$	0.1484	0.1484	0.1483	0.1505	0.1485	−0.5753
$M_{19}[10^{18}]$	−0.0520	−0.0520	−0.0503	−0.0552	−0.0519	0.0603
$M_{20}[10^{18}]$	−0.2045	−0.2044	−0.2058	−0.2041	−0.2044	0.5220
$M_{21}[10^{18}]$	0.1186	0.1186	0.1191	0.1229	0.1185	−0.1292
$M_{22}[10^{18}]$	0.7187	0.7186	0.7195	0.7164	0.7187	−2.3919

$$v_{pq} = \frac{\mu_{pq}}{\mu_{00}^w}, \qquad w = \frac{p+q}{2} + 1.$$

To obtain invariants to image rotation

$$x' = x \cdot \cos \alpha + y \cdot \sin \alpha,$$

$$y' = -x \cdot \sin \alpha + y \cdot \cos \alpha$$

we can utilize well-known rotation invariants introduced by Hu [12]. It can be proved that the following four of them are invariant also to motion blur:

$$\Phi_1 = (v_{30} - 3v_{12})^2 + (3v_{21} - v_{03})^2,$$

$$\Phi_2 = (v_{30} + v_{12})^2 + (v_{21} + v_{03})^2,$$

$$\Phi_3 = (v_{30} - 3v_{12})(v_{30} + v_{12})((v_{30} + v_{12})^2 - 3(v_{21} + v_{03})^2)$$
$$+ (3v_{21} - v_{03})(v_{21} + v_{03})(3(v_{30} + v_{12})^2 - (v_{21} + v_{03})^2),$$

$$\Phi_4 = (3v_{21} - v_{03})(v_{30} + v_{12})((v_{30} + v_{12})^2 - 3(v_{21} + v_{03})^2)$$
$$- (v_{30} - 3v_{12})(v_{21} + v_{03})(3(v_{30} + v_{12})^2 - (v_{21} + v_{03})^2).$$

Since linear motion of arbitrary direction can be decomposed into image rotation, horizontal motion and inverse rotation, the features Φ_1, Φ_2, Φ_3 and Φ_4 are invariant also to the motion direction.

Figure 2. Original image (left), horizontal motion blur ($v \cdot t = 50$ pixels) (middle), diagonal motion blur ($v \cdot t = 30$ pixels) (right)

Table 2. The values of invariants of the images in Fig. 2

	(left)	(middle)	(right)
$\Phi_1[10^{-11}]$	5.4915	5.4130	5.4699
$\Phi_2[10^{-12}]$	5.7614	5.7471	5.8700
$\Phi_3[10^{-23}]$	3.8498	3.7067	3.7377
$\Phi_4[10^{-23}]$	9.4973	9.4345	9.8318

The following experiment deals with image blur caused by non-horizontal motion. According to the analysis given above, rotation invariants Φ_1, Φ_2, Φ_3 and Φ_4 were used as the features for image description. Three test images are displayed in Fig. 2: original image, image blurred by horizontal motion ($v \cdot t = 50$ pixels) and image blurred by motion along a diagonal ($v \cdot t = 30$ pixels). Table 2 illustrates the property of invariance of these features.

7. Discussion

Every features which we want to use for image description should have the following basic properties:

- *Invariance:* Two images belonging to the same class should have the same values of the features.
- *Uniqueness:* Two images which do not belong to the same class should have different values of the features. In other words, the features should be object discriminatory.
- *Stability:* If two images are similar in some manner (in l_2 norm for instance), then their feature values should be similar too, and vice versa.
- *Efficiency:* The features should be efficient to compute and store.

In this Section, we will discuss how the motion blur invariants M_i satisfy the above criteria.

Invariance was theoretically and experimentally proved in Sections 3 and 5.

By uniqueness of the feature we understand the ability to distinguish (in the feature space) among "similar" images. The results of the experiment described in Section

5 have shown the sufficiently high discriminability of the portrait photographs. However, from theoretical point of view, the set of invariants $\mathcal{M}_r$ is an incomplete feature system for every r. That means there may exist two different images $f_1(x, y)$ and $f_2(x, y)$ such that

$$M_j^{(f_1)} = M_j^{(f_2)} \qquad j = 0, 1, 2, \ldots, N_r.$$

There are two different kinds of stability to be investigated in case of blur invariants: stability under additive random noise and stability with respect to boundary effect.

So far we have considered noise-free model (1) only. Now let us consider imaging model with additive zero-mean random noise $n(x, y)$:

$$g(x, y) = (f * h)(x, y) + n(x, y). \tag{6}$$

Since the image $g(x, y)$ is then a random field, all its moments and all invariants can be viewed as random variables. It holds

$$E(\mu_{pq}^{(n)}) = E\left(\int_{-\infty}^{\infty} \int_{-\infty}^{\infty} x^p y^q n(x, y)\, dx\, dy\right) = \int_{-\infty}^{\infty} \int_{-\infty}^{\infty} x^p y^q E(n(x, y))\, dx\, dy = 0$$

and

$$E(\mu_{pq}^{(g)}) = E(\mu_{pq}^{(f*h)}) + E(\mu_{pq}^{(n)}) = \mu_{pq}^{(f*h)},$$

where $E(X)$ denotes the mean value of random variable X.

In practice, however, only single image $g(x, y)$ (i.e. only one realization of a random field) is available in most cases. We obtain $\mu_{pq}^{(g)}$ but we are not able to estimate mean values $E(\mu_{pq}^{(g)})$. Since the moments are computed by a summation over the whole image, they are supposed to be affected by additive noise very little. That means $\mu_{pq}^{(g)}$ are supposed to be close to $E(\mu_{pq}^{(g)})$ and we can use directly $\mu_{pq}^{(g)}$ for the computation of the invariants. It was confirmed by the experiments in Section 5 that the invariants evaluated by this technique are sufficiently stable.

Serious problems can be caused by boundary effects. Provided that the size of the original images is $N \times N$ pixels and the size of PSF support is $1 \times L$ pixels, the correct size of the acquired image should be $N \times (N + L - 1)$ pixels. However, in practice the value of L is unknown and original and acquired images are considered to have the same size. Due to this fact, we lose some valuable information on image boundary. If $L \ll N$, the errors of invariant calculation caused by boundary effect are negligible. If L is relatively large (this comes in case of very fast motion), boundary effect should be taken into consideration. Stability of the invariants with respect to boundary effect should be studied in detail in the near future.

Computing complexity of the motion blur invariants is determined by computing complexity of central moments μ_{pq}. Direct evaluation of μ_{pq} by means of relation (5) requires $O(N^2)$ operations. Although several methods for fast moment computing have been published recently (see [13] for a survey), they are not applicable in case of gray-level images.

J. Flusser et al.

8. Summary and Conclusion

The paper was devoted to the construction of image features invariant with respect to motion blur. The images degraded by relative motion of the object and the sensor can be described as an output of a linear shift-invariant imaging system, whose point spread function has the form (2).

A set of motion blur invariants based on image moments was introduced in this paper. An original algorithm for invariants construction was presented and some invariants were shown in explicit form. The derivation of the invariants is a major theoretical result of the paper.

Invariance of the features as well as their ability to distinguish among different images were demonstrated experimentally. Stability of the invariants with respect to additive random noise was also discussed and was shown to be sufficiently high.

Although the presented method for blurred image description produces very good results in some experiments, it has, however, several limitations. The most significant one arises from the fact that the whole image must be degraded uniformly. This assumption holds good for instance in case of a flat static scene and moving sensor. On the other side, if we can a small moving object in front of a static background, our method is inapplicable.

References

[1] Pratt, W. K.: Digital image processing, 2nd ed. New York: J. Wiley 1991.
[2] Gennery, D. B.: Determination of optical transfer function by inspection of frequency-domain plot. J. Opt. Soc. Amer. *63*, 1571–1577 (1973).
[3] Stockham, T. G., Jr., Cannon, T. M., Ingebretsen, R. B.: Blind deconvolution through digital signal processing. Proc. IEEE *63*, 678–692 (1975).
[4] Chang, M. M., Tekalp, A. M., Erdem, A. T.: Blur identification using the bispectrum. IEEE Trans. Acoust. Speech Signal Proc. *39*, 2323–2325 (1991).
[5] Cannon, T. M.: Blind deconvolution of spatially invariant image blurs with phase. IEEE Trans. Acoust. Speech Signal Proc. *24*, 58–63 (1976).
[6] Jain, A. K.: Advances in mathematical models for image processing. Proc. IEEE *69*, 502–528 (1981).
[7] Tekalp, A. M., Kaufman, H., Woods, J. W.: Identification of image and blur parameters for the restoration of noncausal blurs. IEEE Trans. Acoust. Speech Signal Proc. *34*, 963–972 (1986).
[8] Lagendijk, R. L., Biemond, J., Boekee, D. E.: Identification and restoration of noisy blurred images using the expectation-maximization algorithm. IEEE Trans. Acoust. Speech Signal Proc. *38*, 1180–1191 (1990).
[9] Reeves, S. J., Mersereau, R. M.: Blur identification by the method of generalized cross-validation. IEEE Trans. Image Proc. *1*, 301–311 (1992).
[10] Savakis, A. E., Trussel, H. J.: Blur identification by residual spectral matching. IEEE Trans. Image Proc. *2*, 141–151 (1993).
[11] Andrews, H. C., Hunt, B. R.: Digital image restoration. Englewood Cliffs: Prentice-Hall 1977.
[12] Hu, M. K.: Visual pattern recognition by moment invariants. IRE Trans. Inf. Theory *8*, 179–187 (1962).

[13] Reiss, T. H.: Recognizing planar objects using invariant image features. Lecture Notes in Computer Science, Vol. 676. Berlin Heidelberg New York Tokyo: Springer 1993.

J. Flusser, T. Suk, S. Saic
Institute of Information Theory and Automation
Academy of Sciences of the Czech Republic
Pod vodárenskou věží 4
18208 Prague 8
Czech Republic
e-mail: flusser@utia.cas.cz

Computing Suppl. 11, 53–71 (1996)

Symmetric Bi- and Trinocular Stereo: Tradeoffs between Theoretical Foundations and Heuristics

G. L. Gimel'farb, Budapest

Abstract

Symmetric Bi- and Trinocular Stereo: Tradeoffs between Theoretical Foundations and Heuristics. Tradeoffs between theoretical and heuristic sides of ill-posed problems of the intensity-based computational stereo are discussed, as applied to the previously proposed symmetric approach for solving this problem. The heuristics are needed to deal with discontinuities in stereo images due to partial occlusions of observed surface. Basically, it is these discontinuities that cause the ill-posedness of the stereo problems. Theoretical base of the symmetric stereo is refined here by introducing a novel probabilistic model of the surface geometry and by deducing compound Bayesian decision rules to be implemented by dynamic programming techniques, as in the case of simple MAP-decision with the maximum *a posteriori* probability. Also, several heuristics are proposed to regularize the binocular stereo.

Key words: Computational stereo, ill-posedness, regularization.

1. Introduction

Computational stereo is of great interest in many applications: photogrammetry and mapping, remote sensing of the Earth's surface, industrial (robotics) vision, and automatic navigation are among them. Theoretical and experimental investigations of this problem extend over a protracted period of several tens of years and resulted in numerous approaches to state and solve it (many of them were reviewed and discussed in [1–3], in particular, the intensity-based, edge-based, feature-based approaches, etc.).

Intensity-based stereo can be thought as the lowermost level of the computational stereo vision because a desired optical surface is reconstructed directly by measuring similarities between optical signals, or intensities in initial stereo images of the surface. Most often, initial data consist of a single stereo pair or triple and leads to a so-called *static intensity-based stereo*.

The static stereo belongs to the domain of *ill-posed inverse photometric problems* because of principal multiplicity of optical 3D surfaces giving the same stereo pair or triple. Thus, it is impossible to reconstruct precisely the real surface giving these stereo images. Nonetheless, by using appropriate theoretical models of the stereo images and optical surfaces and proper heuristics the reconstructed surface can be brought close enough to the one perceived visually from a single stereo pair or triple.

Typically, the theoretical models relate optical signals (or *intensities*) in the bi- or trinocularly visible surface points and in the corresponding pixels of the stereo images. Using the model, a measure of similarity between corresponding continuous regions in the images can be deduced. The measure is used to reconstruct a continuous patch of the surface depicted in these image regions. However, some heuristics are needed to deal with possible discontinuities in the stereo images. The discontinuities are due to partial occlusions of the surface with respect to the stereo viewing. The occlusions lead to only a monocular visibility of certain surface points.

1.1. Symmetric Intensity-Based Stereo

Here we restrict the consideration to a *symmetric intensity-based static stereo*. This approach was proposed in [4] and the ensuing developments of it are presented in [5, 6]. It was the first to use *generating probability models* of the stereo images and surface and *Bayesian decision rules* to reduce the problem of the static stereo to a non-linear optimization problem solved by a *dynamic programming* (DP) technique. The introduced model relates geometric and radiometric features of the surface and stereo images with due account of symmetries between optical channels and image sensors and of geometric and radiometric image distortions typical to the stereo observation of the surface.

The geometric model describes one-to-one relations between spatial (3D) Cartesian coordinates of the surface points and planar (2D) coordinates of the corresponding pixels in the raster stereo images. The surface points which correspond to the pixels form a **digital elevation model** (DEM) of the surface.

The radiometric, or photometric, *model* describes the main interactions between the light intensities, or coloring, in the surface points and the sensed intensities in the corresponding pixels. We consider such interactions as probabilistic in essence and treat the computational stereo as a problem of the statistical pattern recognition. More precisely, it is the image recognition problem—to emphasize the main feature of the initial and desired data: an inherent structure of signal interactions [11] in the images and in the DEM. The generating model can be used to solve either the direct photometric problem of simulating stereo images of an optical surface or the inverse problem of reconstructing the surface from the images.

To simplify the problem, we assume that the DEM is formed as a bunch of *epipolar profiles* obtained by crossing the surface by a fan of base planes. Each base plane contains the base-line connecting optical centers of the channels. The corresponding *epipolar lines* cross the stereo images along the traces of the particular base plane and are the same for any possible epipolar profile within this plane. The DEM is represented either as the dense raster map of x-parallaxes between the corresponding pixels in stereo images or as the like map of derived distances (depths) to surface points. Such maps are usually called *range images* and possess their own internal structure of the pixel interactions.

In [4–6] the DEM reconstruction, in a simplified profile-by-profile mode, is reduced to the simple Bayesian MAP decision that looks for the surface profiles with maximum *a posteriori* probabilities. Each profile is reconstructed by an exhaustive DP search through all the possible profile variants. The variant that maximizes a particular additive measure of similarity between the coloring, estimated for the profile points, and the intensities in the corresponding pixels along the epipolar lines has to be found by this search. The similarity measure is deduced from the probabilistic model of the surface and images with due regard for:

(*i*)   a *geometric symmetry* of viewing conditions in the optical channels and image sensors,

(*ii*) independent random and interdependent regular *distortions* of the initial stereo images, and

(*iii*) *discontinuities* in the images because of possible occlusions of the surface parts in each channel.

The occlusions result in the ill-posedness of this inverse photometric problem [6, 8–10]. A principal ambiguity of the binocular surface reconstruction depends upon the fact that even without the signal distortions two or more variants of the surface always are in full agreement with the same stereo pair. For instance, in the binocular case the "pit-like" profile with only the monocularly visible points can give under the proper surface coloring exactly the same stereo pair as any other possible variant with both monocularly and binocularly visible points [8].

1.2 Theoretical and Heuristic Sides of the Symmetric Stereo

We discuss below the tradeoffs between theoretically justified and heuristic sides of the symmetric approach. The introduced heuristics have to reduce the principal multiplicity of the surface variants giving the same high similarity between the image intensities and surface coloring. Of course, no theoretical models or heuristics can overcome this multiplicity and reconstruct precisely the surface which actually was observed and originated the initial stereo images. But, by adjusting the models and heuristics one can bring the reconstructed surface close enough to the one perceived visually from the single stereo pair or triple and approach the accuracy that human stereo vision has under this very restrictive condition of only a single (static) observation of the surface. In practice, the static stereo is most typical for the aerial and space remote sensing, stereophotogrammetry, and mapping.

The theoretical foundations of the symmetric approach (see [4–6]) are amplified here by introducing: (*i*) compound Bayesian decisions which are generally more appropriate [5, 9, 11, 12] to reconstruct the raster DEMs than the traditional simple MAP-decision, (*ii*) a symmetric geometric framework for both the binocular and trinocular stereo unifying the DP implementations of the simple MAP-decision and compound Bayesian ones, and (*iii*) a generative probability model of the epipolar profile that allows to describe roughly a geometric shape of the desired profiles. Also, we propose and discuss some heuristics that help to overcome partly the intrinsic ambiguity of the binocular static stereo.

The paper is organized as follows. In Section 2 the unified geometric scheme for the symmetric bi- and trinocular vision is described and compound Bayesian decisions for reconstructing the epipolar profiles are deduced in the general form. In Section 3 we discuss a probability model of the profile proposed in [7] to describe the local smoothness of the profiles and improve the ordering of profile variants in the DP reconstruction. The DP algorithms and possible heuristics to regularize them are considered in Section 4. A few experimental results are presented in Section 5 to confirm a feasibility of the introduced theoretical model and regularizing heuristics.

2. Unified Bayesian Framework for Bi- and Trinocular Symmetric Stereo

A joint epipolar geometric model of the stereo observation for both the symmetric binocular and trinocular cases is presented in this section, and compound Bayesian decision rules to reconstruct the DEM from the given stereo images are deduced in the general form.

The map of x-parallaxes is subject to tight constraints on possible changes between the adjacent parallaxes along the epipolar lines. These constraints result in deriving the compound rules which are not reduced to a known point-wise MPM-decision with maximal marginal posterior probabilities of the surface points [9] and MPE-one with marginal posterior expectations to be chosen as these points [12]. The obtained compound rules involve a maximization of a particular additive measure of the signal similarity or minimization the like dissimilarity measure.

So, together with the simple MAP rule, they form an unified Bayesian framework for the symmetric bi- and trinocular static stereo. As will be shown in Section 4, this framework leads to rather similar DP implementations of these Bayesian decision rules for reconstructing the epipolar profiles of the surface.

2.1. Epipolar Geometry of the Symmetric Stereo

We will use the following notation to describe the raster (digital) geometric and radiometric models of the stereo images and DEM to be reconstructed. Let R be a finite 2D arithmetic lattice, or raster, that supports the DEM and $Rj; j \in J$ denote the like rasters supporting the stereo images ($J = \{1, 2\}$ in the binocular and $J =$

$\{0, 1, 2\}$ in the trinocular case):

$$R = ((x, y): x \in X = \{x_b, \dots, x_e\}; y \in Y = \{y_b, \dots, y_e\});$$

$$Rj = ((xj, yj): xj \in Xj = \{xj_b, \dots, xj_e\}; yj \in Yj = \{yj_b, \dots, yj_e\}).$$

All the rasters have the epipolar x-lines. We assume that the DEM is reduced to the plane of the stereo images and composed from the epipolar profiles aligned with the x-axis of the Cartesian coordinates [6]. The rasters R and $R0$ have x-steps of 0.5 and y-steps of 1:

$$x = x_b, x_b + 0.5, x_b + 1, \dots, x_e - 0.5, x_e;$$

$$x0 = x0_b, x0_b + 0.5, x0_b + 1, \dots, x0_e - 0.5, x0_e;$$

$$y = y_b, y_b + 1, \dots, y_e - 1, y_e; y0 = y0_b, y0_b + 1, \dots, y0_e - 1, y0_e.$$

the rasters Rj with $j = 1, 2$ have x- and y-steps of 1:

$$xj = xj_b, xj_b + 1, \dots, xj_e - 1, xj_e; yj = yj_b, yj_b + 1, \dots, yj_e - 1, yj_e$$

where symbols "b" and "e" denote the beginning and ending values in each direction.

The symmetric stereo system has identical optical channels with parallel optical axes and identical image sensors that form the ideal horizontal stereo pair or triple. In such a system the raster coordinates are related by the following geometric model [4, 6]:

$$x0 = x; x1 = x + p/2; x2 = x - p/2; y0 = y1 = y2 = y. \tag{1}$$

Here $p = x1 - x2$ is the x-parallax for the pixels $(xj, yj); j \in J$ that correspond in the stereo images to the surface point with planar (raster) coordinates (x, y).

Figure 1 shows an epipolar cross-section of the stereo channels both in the symmetric bi- and trinocular cases. The x-parallax p is inversely proportional to the depth (distance, or height) Z of the surface point (X, Y, Z) from the focal plane OXY: $p = B * F/Z$ where B denotes the length of base-line of the stereo channels, F is the focal length having the same value in all the channels, and X, Y, Z denote symmetric 3D Cartesian coordinates of the visible surface points. The base-line links optical centers $O1$ and $O2$ of the channels 1 and 2; the third center O in the symmetric trinocular case is midway between them and coincides with the origin of the 3D coordinates [6].

Lines $o1x1$, $ox0$, and $o2x2$ in Fig. 1 represent, respectively, the x-axes of the rasters $R1$, $R0$, $R2$. The digital profile in the base plane OXZ is a chain of isolated spatial points having the corresponding image pixels with integer x-coordinates $x1$ and $x2$ (see Fig. 1).

The geometric description of the DEM to be reconstructed from the digital stereo images is simplified by projecting these spatial points onto the raster $R0 \equiv R$ in

G. L. Gimel'farb

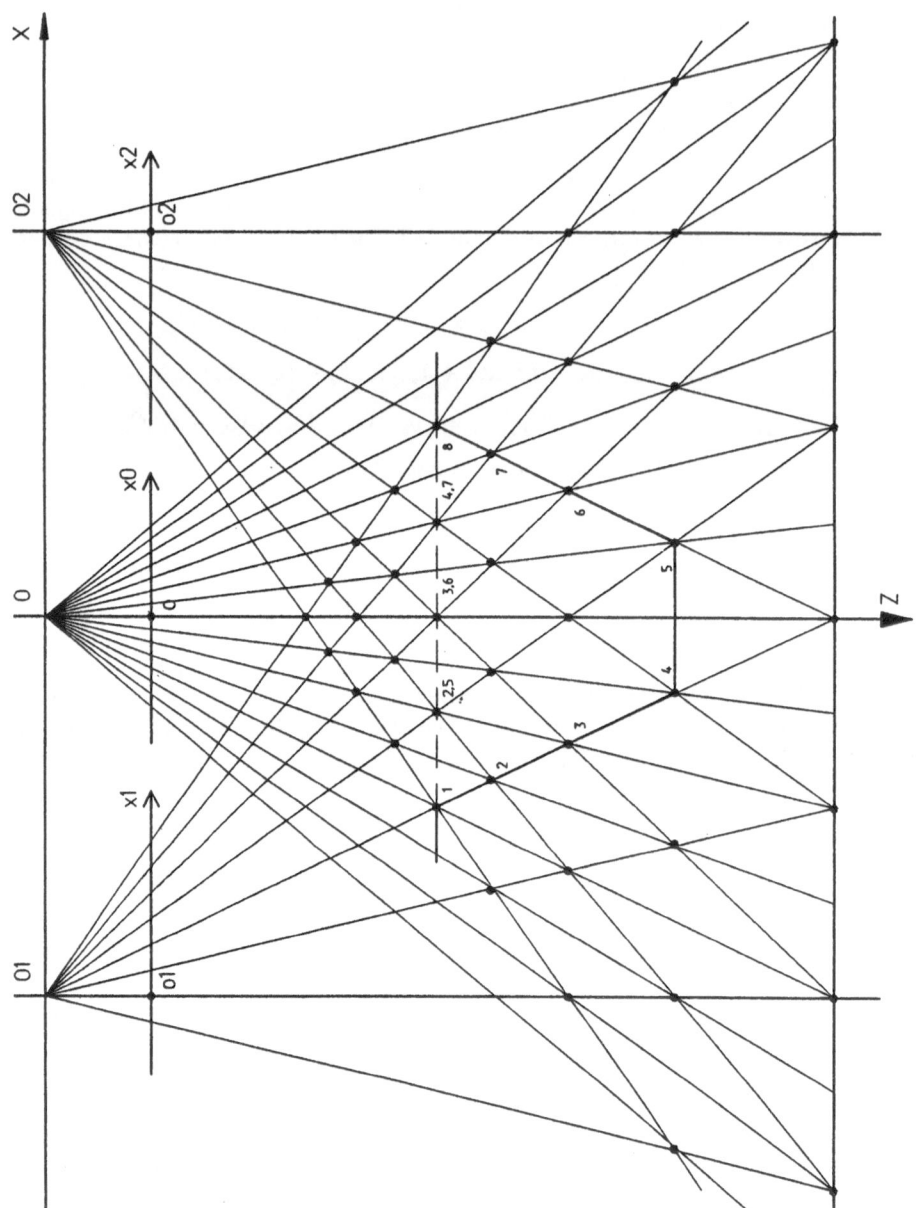

Figure 1. Epipolar geometry of the symmetric bi- and trinocular stereo

the image plane and replacing z-coordinates of the surface points with the x-parallaxes. The DEM (TG, TI) consists of two parts: the *geometric part TG* represents the *digital surface* itself and the *radiometric part TI* is the *orthoimage* of the surface showing the coloring, or light intensities of the surface points. The given images $(Cj: j \in J)$ and the desired DEM can be considered as functions on the supports R, Rj [5, 6]:

$$TG: R \rightarrow P; \quad TI: R \rightarrow V; \quad Cj: Rj \rightarrow V; \quad j \in J. \qquad (2)$$

Here, P denotes a finite set of the possible x-parallax values and V is a finite set of the possible intensity values in the pixels and surface points.

Any epipolar profile $EPy: X \rightarrow P$ has a fixed y-coordinate of the points in the surface TG represented as a bunch of the profiles: $TG = (EPy: y \in Y)$. Each profile EPy corresponds to the epipolar lines $Cj_y \equiv cj: Xj \rightarrow V; j \in J$, in the stereo images. These lines have the same y-coordinate $yj = y$ of the pixels.

2.2. Simple and Compound Bayesian Decisions in Computational Stereo

Let us assume that the probability models of the surface and stereo images give *a posteriori* probability distribution (p.d.) $\Pr(TG|Cj: j \in J)$ of the desired surface under the given images. Then the usual simple Bayesian MAP-decision reconstructing the surface with the maximum *a posteriori* probability can be deduced [5, 6, 9–12]:

$$TG^{opt} = \arg\max_{TG} \Pr(TG|Cj: j \in J). \qquad (3)$$

This decision rule minimizes the error probability if the error is considered as any discrepancy between the true and obtained surface regardless of a number and magnitudes of pointwise errors. The *pointwise error* can be defined as the difference between true and reconstructed x-parallax values for a particular surface point. The compound Bayesian decision rules are more adequate to the low-level stereo because take into account the features of the pointwise errors [5, 9]. The usual compound rules minimize either the expected number of the pointwise errors or the expected sum of the squared errors (that is, the variance of the obtained DEM about the true one) and can be written as follows:

$$TG^{opt} = \arg\max_{TG} \sum_{(x,y) \in R} \mathrm{MPr}_{x,y}(p = TG(x, y)|Cj: j \in J); \qquad (4)$$

$$TG^{opt} = \arg\max_{TG} \sum_{(x,y) \in R} (TG(x, y) - MPE(x, y|Cj: j \in J))^2. \qquad (5)$$

Here, $MPE(\ldots)$ denotes the surface formed by the marginal posterior expectations of the x-parallax value for the points $(x, y) \in R$:

$$MPE(x, y|Cj: j \in J) = \sum_{p \in P} p * \mathrm{MPr}_{x,y}(p|Cj : j \in J) \qquad (6)$$

and $\mathrm{MPr}_{x,y}(p|Cj: j \in J)$ denotes the marginal posterior probability of the surface point $(x, y, p = TG(x, y))$ under the given images.

The x-parallaxes in the neighboring points along the epipolar profile in the DEM are subject to inherent tight constraints dictated by the visibility conditions [4–6]:

$$|TG(x, y) - TG(x - 0.5, y)| \leq 1; \quad |TG(x, y) - TG(x - 1, y)| \leq 2. \tag{7}$$

They prevent to reduce the compound MSMPP-decision in Eq. (4) having maximal sum of the marginal posterior probabilities of the profile points and MVMPE-one in Eq. (5) with minimal variance of these points about the marginal posterior expectations, respectively, to the known and used in [5, 6, 9, 11, 12] pointwise MPM-decision with the maximal marginal posterior probability of each profile point and MPE-decision with the posterior expectations as these points. So, the latter decision rules will not do for solving the stereo problems and should be replaced by the MSMPP- and MVMPE-decisions in Eqs. (4) and (5).

3. Markov Chain Models of Epipolar Profiles

In this section we describe a probabilistic model of an epipolar profile introduced in [7]. The model represents the profile as a sample of a Markov chain of admissible transitions between successive points in a symmetric planar *graph of variants of the profile* (GVP). The GVP-points represent the surface points for all the possible profiles in the coordinates x, p. For the binocular case such a GVP has been proposed in [4, 6]. Each GVP-point has three states indicating tri-, bi-, or only monocular visibility of the corresponding surface point. The Markov chain is assumed to have a stationary probability distribution of the visibility states. The model allows to describe expected shapes and smoothness of the profiles by using transition and marginal probabilities of these states. Thus, it introduces an additional probabilistic ordering of the profile variants having the same similarity between the image intensities and surface coloring.

3.1. A Priori Probability Model of Epipolar Profiles

Let us assume that local profile slopes are bounded by the maximal slopes for the binocular stereo. Then the symmetric bi- and trinocular cases differ only in particular visibility states s of any GVP-point (x, p, s): $s \in \{B, M1, M2\}$ indicating binocular (B) or only monocular (M1, M2) observation by the channel 1 or 2 in the first case and $s \in \{T, B10, B02\}$ indicating trinocular (T) and binocular observation (B10, B02) by the channels 1 and 0 or 0 and 2 in the second case (Fig. 2).

Let $\pi(x_c, p_c, s_c | x_p, p_p, s_p)$ be a probability of transition to the GVP-point (x_c, p_c, s_c) from the GVP-point (x_p, p_p, s_p). There are eight possible transitions between the current point (x_c, p_c, s_c) and its nearest preceding neighbors (x_p, p_p, s_p) along the profile [5, 6] (see Fig. 2). It is obvious that the differences $x_c - x_p$ and $p_c - p_p$ between their x-coordinates and parallaxes are specified uniquely for any transition by the visibility states (s_c, s_p). So, the transition probabilities can be denoted as $\pi(s_c | s_p)$. Only seven of the allowable transitions have non-zero probability if the

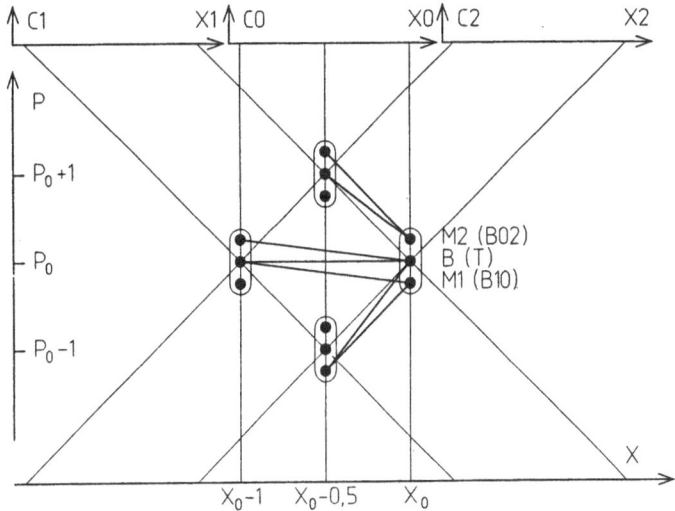

Figure 2. Admissible transitions in the GVP with regard to the visibility conditions

considered Markov chain have the stationary p.d. of the visibility states in the equilibrium. As shown in [7], this case leads to the zeroth probability of the transition M2 → M1:

$$\pi(M1|M2) \equiv \pi(x_c, p_c, s_c = M1 | x_p, p_p, s_p = M2) = 0.$$

An example of the resulting GVP is shown in Fig. 3. A very narrow "tube" of six possible parallax values is used here to illustrate the uniformity of all transitions except for the uppermost and lowermost ones. Let M substitutes for M1 or M2. Then, for the given transition probabilities $\pi(B|B)$ and $\pi(M|M)$, the following relationships between the transition probabilities $\pi(s_c|s_p)$ to generate the profiles in the GVP and resulting marginal probabilities of the visibility states MPr(B) and MPr(M) hold:

$$\pi(M|B) = (1 - \pi(B|B))/2; \quad \pi(B|M) = 1 - \pi(M|M); \quad \pi^0(M|M) = 0;$$
$$\text{MPr}(B) = \xi/(\xi + 1); \quad \text{MPr}(M) = 0.5/(\xi + 1). \tag{8}$$

Here, $\xi = (1 - \pi(M|M))/(1 - \pi(B|B))$, $\pi(M|B) \equiv \pi(M1|B) = \pi(M2|B)$, $\pi(B|M) \equiv \pi(B|M1) = \pi(B|M2)$, $\pi(M|M) \equiv \pi(M1|M1) = \pi(M2|M2)$, and $\pi^0(M|M) \equiv \pi(M1|M2)$.

To retain the equilibrium conditions at the uppermost and lowermost boundaries of the GVP (see Fig. 3), we need only to introduce the following specific transition probabilities for the extreme GVP-points at the uppermost boundary:

$$\pi_{upp}(B|B) = 1 - \text{MPr}(M); \quad \pi_{upp}(B|M) = \text{MPr}(M); \quad \pi_{upp}(M|B) = 1$$

and at the lowermost boundary:

G. L. Gimel'farb

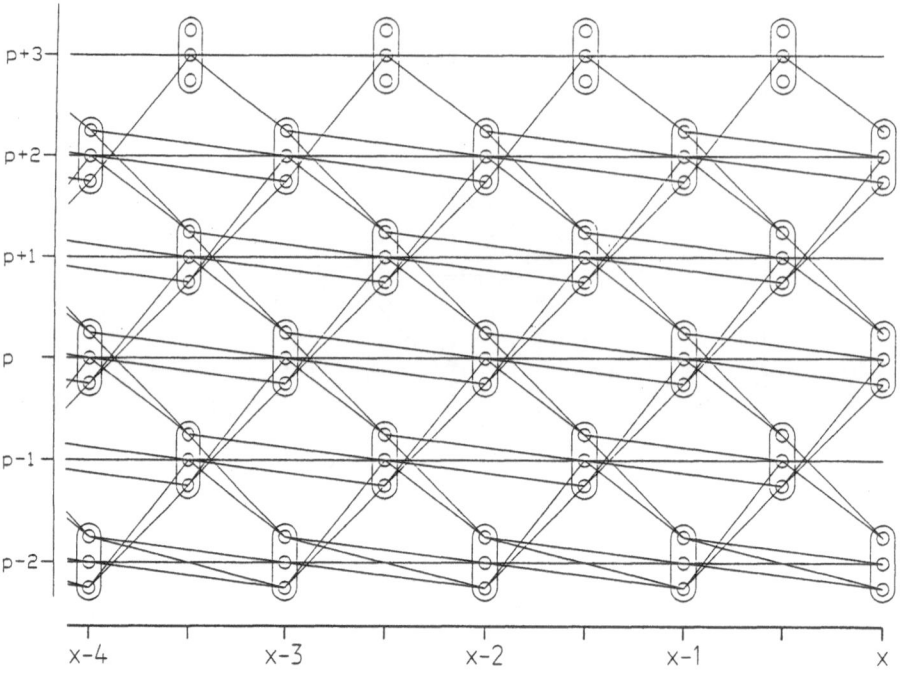

Figure 3. An example of the GVP for the symmetric bi- and trinocular stereo

$$\pi_{\mathrm{low}}(B|B) = 1 - \kappa; \quad \pi_{\mathrm{low}}(B|M) = \kappa; \quad \pi_{\mathrm{low}}(M|B) = 1; \quad \pi_{\mathrm{low}}^0 = 0 \quad \text{if } \kappa \le 1;$$

$$\pi_{\mathrm{low}}(B|B) = 0; \quad \pi_{\mathrm{low}}(B|M) = 1; \quad \pi_{\mathrm{low}}(M|B) = 1/\kappa; \quad \pi_{\mathrm{low}}^0 = (\kappa - 1)/\kappa \quad \text{if } \kappa > 1$$

where $\pi_{\mathrm{low}}^0(M|M) \equiv \pi(M1|M2); \kappa = \mathrm{MPr}(M)/\mathrm{MPr}(B)$.

3.2. A Posteriori Probability Model of Epipolar Profiles

To unify the DP reconstruction of the profiles described by this GVP let us assume that the *a posteriori* p.d. $\mathrm{Pr}(EP|cj: j \in J)$ for the profile $EP \equiv EP(y): X \to P$ within the GVP in a rather general case can be factored into a product of conditional transition probabilities. Here, $cj \equiv Cj(y) = (Cj(xj, yj): xj \in Xj; yj = y = const)$ are the epipolar lines in the images that correspond to the epipolar profile EP with the fixed y-coordinate $y = const$. These transition probabilities take account of the geometric and radiometric features of the desired profile EP and of the radiometric features of the initial image intensities $cj; j \in J$, to relate the similarity of the intensities under the given profile to the *a posteriori* probability of the profile. The *a posteriori* probability is used to obtain the Bayesian decision rules in Eqs. (3)–(5) for the reconstruction of the desired profiles [4–6]:

$$\mathrm{Pr}(EP|cj: j \in J) = \mathrm{MPr}_0(x_0, p_0, s_0|cj: j \in J) *$$
$$\prod_{i=1}^{N-1} \mathrm{TPr}(x_i, p_i, s_i|x_{i-1}, p_{i-1}, s_{i-1}; cj: j \in J). \tag{9}$$

Here, i denotes serial numbers of the GVP-points along the profile, N is the total number of these points, $MPr_0(x_0, p_0, s_0 | cj : j \in J)$ denotes the marginal probability of the starting GVP-point (x_0, p_0, s_0) in the profile variant under the given image intensities $cj; j \in J$, and $TPr(x_c, p_c, s_c | x_p, p_p, s_p; cj : j \in J)$ denotes the probability of the transition between the successive GVP-points (x_p, p_p, s_p) and (x_c, p_c, s_c) along the profile variant under the same condition. The transition probabilities are derived using the probabilistic geometric models like that given in Eq. (8) and radiometric models of the admissible variations of the intensities in the corresponding pixels as against the coloring in the surface points.

3.3. Marginal Probabilities of the GVP-Points

The marginal probabilities in Eq. (9) can be calculated consecutively along the GVP by the obvious relations which follow directly from Figs. 2 and 3:

$$Mpr(x_c, p_c, s_c | cj : j \in J) = \sum_{\substack{x_p, p_p, s_p \in \\ \omega(x_c, p_c, s_c)}} MPr(x_p, p_p, s_p | cj : j \in J) *$$

$$TPr(x_c, p_c, s_c | x_p, p_p, s_p; cj : j \in J). \tag{10}$$

Here $\omega(x_c, p_c, s_c)$ denotes the set of the nearest neighboring GVP-points (x_p, p_p, s_p) that precede the current GVP-point (x_c, p_c, s_c). Generally it contains the following points (Fig. 2):

$$\omega(x_c, p_c, M1) = \{(x_c - 1, p_c, B), (x_c - 0.5, p_c - 1, M1)\};$$

$$\omega(x_c, p_c, B) = \{(x_c - 1, p_c, B), (x_c - 1, p_c, M2), (x_c - 0.5, p_c - 1, M1)\}; \tag{11}$$

$$\omega(x_c, p_c, M2) = \{(x_c - 0.5, p_c + 1, B), (x_c - 0.5, p_c + 1, M2)\}.$$

The relation in Eq. (9) leads to an additive similarity measure based on logarithms of the transition probabilities $TPr(\ldots)$ along the profiles within the GVP to implement the simple Bayesian MAP-decision rule of Eq. (3). The relations in Eqs. (10) and (11) allow to compute the additive similarity measures of Eqs. (4) and (5) for the compound Bayesian rules that are based on the marginal posterior probabilities $MPr(\ldots)$ of the GVP-points.

4. Dynamic Programming Reconstruction of Epipolar Profiles

The above relations in Eqs. (9) and (10) allow to implement the Bayesian profile reconstruction by using the DP techniques to maximize the resulting additive similarity measures or minimize the like dissimilarity ones [4–6]. In this section we consider such DP algorithms and discuss possible heuristic regularizations of them.

The symmetric trinocular stereo presents one of the possible ways to overcome the ambiguity of the surface reconstruction if the surface slopes vary within the limits such that any surface point is observed at least binocularly. In the binocular stereo these limits correspond to occlusions and result in the monocular observation of the occluded surface points. In the symmetric trinocular stereo, as shown in Fig. 2, the multiple solutions are obtained mostly for the surface patches with the constant (uniform) coloring because any GVP-point is observed at least binocularly.

But the static binocular stereo itself, without some suitable heuristics, cannot cope with the discontinuities in the images due to the occlusions and with resulting multiplicity of the surfaces that are consistent with the images. The heuristics are needed, in particular, to estimate the coloring in and define the signal similarity for the monocularly visible points of the surface. Such the heuristics have to be used along with the probabilistic generating signal models and the decision rules. In this section we describe a similarity measure for the stereo images proposed in [7] to obtain more close correspondence between the reconstructed DEM and visually perceived surfaces. The measure contains a weighted linear combination of two similarities: (*i*) between the intensities in the rectified stereo images and the estimated surface coloring and (*ii*) between both rectified stereo images in themselves only. The rectification reduces each stereo image onto the surface (that is, into an orthoimage of the surface) in accord with the obtained DEM. The first similarity measure has been deduced in [4–6] from the generating model of the surfaces and stereo pairs that takes account of the possible discontinuities due to the occlusions. The second measure is deduced from the like generating model, but describing only mutual distinctions between the images in the stereo pair. As a result, the profile variants can be ordered not only in the similarity between the stereo images and surface coloring to within the admissible geometric and radiometric distortions (as in the known algorithms [4–6]) but also in the similarity between the superposed rectified images. Under the equal similarity between the images and surface coloring such heuristic ordering allows to offer the variant with the minimum number of the occlusions.

4.1. Difference Radiometric Model to Measure Similarity of Stereo Images

In the general case the symmetric radiometric model of the stereo images $Cj; j \in J$, is introduced by specifying their distortions with respect to the surface coloring TI. We can describe the distortions by positive *transfer factors* $TFj; j \in J$, that vary over a field-of-view (FOV) of each stereo channel and by a *random noise* $RNj; j \in J$, in the channels:

$$vj = aj * V + rj; \quad j \in J. \tag{12}$$

Here, $V = TI(x, y)$ denotes the coloring value $V \in V$ in the point $(x, y, z = TG(x, y))$ of the surface (TG, TI); $vj = Cj(xj, y)$ is the intensity value $vj \in V$ in the corresponding pixel of the stereo image Cj; $aj = TFj(xj, y)$ denote the transfer factor for this signal that describes a multiplicative part of the distortions and varies in the given range $aj \in A = [a_{\min}, a_{\max}]; 0 < a_{\min} \le a_{\max}$; and $rj = RNj(xj, y)$ is the random noise that describes an additive part of the distortions. By introducing certain assumptions about the regular and random interdependencies between the transfer factors over the FOVs of the channels and about the p.d. of the additive noise one can deduce different statistical estimates for the surface coloring and transfer factors. These estimates, obtained from the given image intensities in the pixels that correspond to the bi- (BVP) or trinocularly visible surface points (TVP), form a theoretically justified part of the quantitative intensity-based measure of similarity or dissimilarity between the stereo images under the given surface geometry TG (see [4–6] for more details).

The heuristic part of the measure corresponds to monocularly visible points (MVP) of the surface because the model in Eq. (12) gives no ways to estimate the introduced parameters of the intensity distortions without prior assumptions about their links with the like parameters for the neighboring BVPs or TVPs.

The transfer factors TFj represent the most regular part of the image distortions with respect to the surface coloring. So, they must have constrained variations over the adjacent BVPs and TVPs for retaining the visual resemblance between the stereo images. This salient feature of the images can be described, in particular, by the symmetric difference radiometric model proposed in [4–6]. This model assumes the direct proportion between each coloring change in the adjacent BVPs and the corresponding intensity changes in the stereo images to within the additive random noise:

$$aj * V - aj' * V' = ej * (V - V'); \quad j \in J. \tag{13}$$

Here $V = TI(x, y)$ and $V' = TI(x', y')$ are the coloring values in the neighboring BVPs $(x, p = EPy(x), B)$ and $(x', p' = EPy'(x'), B)$ along the same epipolar profile $(y = y')$ or in two adjacent ones $(y \neq y')$ in the DEM $TG = (EPy: y \in Y)$ and $ej \in E$ denotes positive "difference" transfer factors varying within the given range $E = [e_{min}, e_{max}]; 0 < e_{min} \leq e_{max}$. These factors describe local interdependencies between the "amplitude" factors TFj over the FOVs. The difference model in Eq. (13) admits large deviations between the corresponding intensities but retains the visual resemblance of the images by preserving approximate direct proportions between their changes.

Under this model the similarity between the images is specified by the transitional probabilities in Eq. (9) and can be derived by taking account of maximal residual deviations of each rectified stereo image from the estimated surface coloring. To obtain these deviations, the estimated coloring is transformed to within the given range E of the admissible difference transfer factors to find the best approximation of each stereo image [4–6]. For simplicity, let us consider a single epipolar profile EPy in the binocular case and denote vj_i the intensities in the pixels $(xj_i, yj_i = y)$; $j \in J = \{1, 2\}$, corresponding to the GVP-point $(x_i, p_i, s_i = B)$ with an ordinal number i along the profile. Let us denote also val_i the intensity approximating vl_i when the estimated coloring of the surface is transformed as to minimize the maximal square error for both the images in accord with the given range E of difference transfer factors. Then under a particular p.d. of the random noise $RNj; j = 1, 2$, the transition probabilities for the neighboring BVPs can be presented as follows (see [5–7] for more details):

$$\text{TPr}(x_i, p_i, s_i = B | x_{i-1}, p_{i-1}, s_{i-1} = B; cj: j = 1, 2) = \text{const}_{i,i-1} * \pi(B|B) *$$
$$\exp(-\gamma * d(vj_i, vj_{i-1}: j = 1, 2)). \tag{14}$$

Here, the factor γ is proportional to an expected variance of the residual minimax error $\delta 1_i = v1_i - val_i \equiv -\delta 2_i$ of the approximation. The local dissimilarity measure $d(\dots) = (\delta 1_i)^2$ depends on the corresponding differences (intensity changes) $\Delta j_{j,i-1} = vj_i - vj_{i-1}$ in both images, on the current estimate of the difference transfer factors for these pixels, and on the residual errors as follows:

$$\delta 1_i = \delta 1_{i-1} + \Delta 1_{i,i-1} - e^{\text{opt}} * (\Delta 1_{i,i-1} + \Delta 2_{i,i-1});$$
$$e^{\text{opt}} = \arg \min_{e \in E} (\delta 1_{i-1} + \Delta 1_{i,i-1} - e * (\Delta 1_{i,i-1} + \Delta 2_{i,i-1}))^2. \tag{15}$$

Here $E = [e_{\min}, e_{\max} = 1 - e_{\min}]; 0 < e_{\min} \leq 0.5$. The transition probability for the transition $M \rightarrow B$ (that is, $s_i = B; s_{i-1} = M$) has the like form except for using the nearest preceding BVP along the profile instead of the adjacent MVP $i - 1$ to get the dissimilarity measure of Eqs. (14) and (15). The relations in Eq. (15) are derived from the model in Eq. (13) by minimizing, in each the surface point, the maximum error of adjusting the estimated surface coloring to the images in the stereo pair. In this case the estimate of the surface coloring v_i for the BVP i is equal to the mean intensity in both pixels: $v_i = (v1_i + v2_i)/2$. The adjustment of the coloring estimates is bounded by the given allowable range E of the difference transfer factors. Note that the transitional probabilities in the trinocular case with only the BVPs and TVPs can be obtained in the similar way.

But in the binocular case the suitable heuristics are essential to cope with $M \rightarrow M$ and $B \rightarrow M$ transitions because for the MVPs we have no like estimates of the transfer factors and surface coloring. Let us assume that variations of the transfer factors and additive noise characteristics in Eq. (12) are rather smooth over the FOVs. Under such an assumption there are several possible ways to introduce the necessary heuristics: (i) a constant value of the residual square error $d(\dots) = d_0 \equiv \delta_0^2 > 0$ for all the MVPs [4–6], (ii) an extension of the absolute error value obtained for the current BVP on the subsequent MVPs [6], (iii) an extension of the relative error of the approximation $\lambda 1_i = \delta 1_i / v1_i$ for the BVP on the subsequent MVPs: $\delta 1_k = \max\{\delta_0, \lambda 1_i * v1_k\}; k = i + 1, i + 2, \dots$, as long as $s = M1$ or $M2$, and so forth.

4.2. Unified DP Framework for the Bi- and Trinocular Stereo

The Markov model in Eq. (9) that considers the profile as the sample of the Markov chain of the transitions between the successive GVP-points, the difference radiometric model leading to the transition probabilities of Eq. (14), and the criteria in Eqs. (3)–(5) allow to unify the Bayesian computational framework for the static stereo. The desired epipolar profile $EP \equiv EPy$ can be reconstructed by means of DP because any above criterion leads to a certain additive similarity (dissimilarity) measure to be maximized (minimized) for the reconstruction:

$$EP^{\text{opt}} = \arg \max_{EP} \text{Sim} \, 1(EP|cj: j \in J). \tag{16}$$

The similarity measure Sim 1(...) can be represented as follows:

$$\text{Sim} \, 1(EP|cj: j \in J) = \sum_{i=1}^{N-1} \text{Loc}(x_i, p_i, s_i; x_{i-1}, p_{i-1}, s_{i-1}|cj: j \in J) \tag{17}$$

where Loc(...) denotes the local similarity term that depends on the transition between the GVP-points $i - 1$ and i.

The difference radiometric model of the stereo images proposed in [4–6] and briefly considered above allows to represent the transition probability TPr(...) as the exponential function with the exponent Loc(...) = log TPr(...) (see Eqs. (14) and (15)). This exponent depends on the intensities and their differences in the stereo images corresponding to the successive GVP-points (x_p, p_p, s_p) and (x_c, p_c, s_c) in the profile and describes quantitatively their visual similarity (or dissimilarity). So, we can get the MAP-variants of the profiles $EP(y)$; $y \in Y$, by searching with the DP for the global maximum of Eq. (3) of the additive similarity measure derived from Eq. (9) or, what is the same, the global minimum of the like dissimilarity measure [4–7]. This DP search involves a successive pass along the x-axis. For any current x-coordinate x_c, all the possible GVP-points (x_c, p_c, s_c) are looked over to calculate and store for each the point the local potentially-optimal backward transition along the profile to one of the preceding points listed in Eq. (11). In more detail this technique is discussed in [4, 6].

By embedding calculations of the corresponding marginals of Eq. (10) into this search we can implement in similar way the compound decisions in Eqs. (4) and (5). It can be shown that the resulting additive similarity measures differ only slightly in the bi- and trinocular cases.

4.3. Heuristics to Regularize DEM Reconstruction

The above similarity measures to reconstruct the DEMs with the DP techniques were deduced from the probabilistic models of the surface and stereo images. But, even being theoretically justified and supplemented by the above heuristics, these measures do not overcome to a full extent the main reasons of inherent ambiguity of the DEM reconstruction. To discuss this ambiguity more in depth let us return to Fig. 1. In the trinocular case one can easily discriminate between the flat profile variant I = ((1)–(2, 5)–(3, 6)–(4, 7)–(8)) and the "pit-like" variant II = ((1)–(2)–(3)–(4)–(5)–(6)–(7)–(8)) by comparing directly the initial intensities under their admissible radiometric distortions (that is, intensities $c1(\ldots)$, $c0(\ldots)$, $c2(\ldots)$ corresponding, respectively, to the successive BVPs (5)–(6)–(7), (2)–(3)–(4)–(5)–(6)–(7), and

(2)–(3)–(4)). But in the binocular case the visually much less appropriate pit-like variant II can concur successfully with I even under the undisturbed intensities. So, we have to introduce the additional heuristics to regularize partly the ill-posed problem of the binocular stereo and obtain resulting DEMs being mostly in line with the visual reconstruction. The heuristics can be based on the expected surface smoothness and admissible distortions of the images with respect to the surface coloring and to each other [5–7].

The heuristic based on the surface smoothness is rather straightforward: under the equal similarity of both the images the most smooth profile (that is, the one with maximal number of the BVPs) has to be chosen by the DP. The less apparent heuristic follows from comparing the orthoimage *TI* obtained by the estimation of the surface coloring with the like orthoimages formed by a separate rectification of each stereo image. This allows to analyze the mismatches between the corresponding intensity changes in the orthoimage and in both the rectified images. This heuristic is based on the assumption that the more deformations (even the admissible ones) are required to transform one stereo image into another, the less probable is the reconstructed surface.

We implement the latter heuristic to circumvent partly the ambiguities in the MAP-decision of Eq. (3) by introducing a weighted sum of two similarity measures to be maximized:

$$EP^{\text{opt}} = \underset{EP}{\arg\max} \left\{ w * \text{Sim } 1(EP|c1, c2) + (1 - w) * \text{Sim } 2(EP|c1, c2) \right\}; \ 0 \le w \le 1.$$

$$(18)$$

Here, Sim $1(EP|c1, c2)$ denotes the measure of similarity between the epipolar lines $c1$ and $c2$ in the stereo images and the estimated surface coloring under the given profile *EP* and Sim $2(EP|c1, c2)$ is the like measure between both images in themselves rectified in accord with the profile. The first measure is deduced from the transition probabilities in Eqs. (14) and (15) for the difference model of the stereo images and surface coloring. This measure allows large variations of the intensities over the images to within the given range of ratios between the corresponding intensity changes for the neighboring BVPs in the surface [4–6]. The second measure estimates the similarity (and thus the visual resemblance) between both the rectified stereo images as a monotone function of the magnitude of deformations that transform one of these images into another. We evaluate this magnitude, in particular, by summing square differences between the corresponding intensity changes along the profile *EP*:

$$\text{Sim } 2(EP|c1, c2) = \sum_{i=1}^{N-1} (\Delta 1_{i, L(i)} - \Delta 2_{i, L(i)})^2. \qquad (19)$$

Here, $\Delta j_{i,k} = cj(x_i) - cj(x_k)$ denotes the intensity change between two pixels along the epipolar line in the rectified stereo image ($k < i$) and $L(i)$ is the last BVP that precedes the GVP-point i along the profile *EP*.

The weighted similarity measure in Eqs. (18) and (19) allows to decide in favour of the profile giving the highest resemblance between both rectified images, all other factors (including the similarity between the images and surface coloring and the surface smoothness) being equal. So, it tries to suppress the pit-like profile variants that cause big differences between the corresponding parts of the rectified images.

The weight w in Eq. (18) can lie in a rather broad range (for instance, the values between 0.2–0.8 gave almost the same betterment of the reconstruction accuracy in our experiments).

5. Experimental Results and Concluding Remarks

Experiments with real stereo pairs indicate that the proposed Bayesian approach gives the dense depth (x-parallax) maps which agree closely with the visually perceived surfaces. To exclude outliers that are due to an inexact epipolar geometry of the initial stereo images or due to a low contrast in the corresponding image regions, the DP reconstruction was followed here by a simple median smoothing between 3–5 adjacent profiles. Figures 4 and 5 present the initial stereo pairs

Figure 4. Reconstruction of the 3D surface from the close-range stereo pair "Plain terrain". Left stereo image (left), right stereo image (middle), range image of the obtained x-parallax map (right)

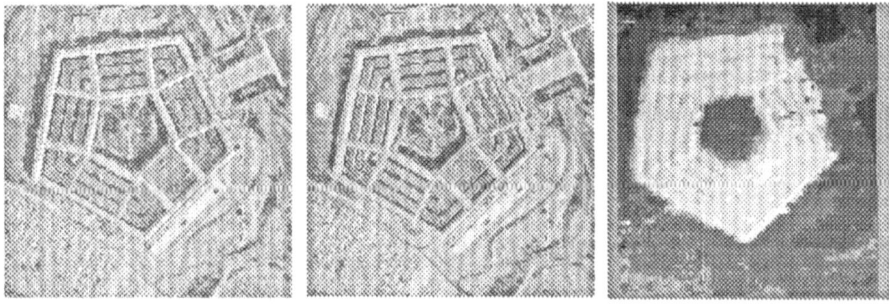

Figure 5. Reconstruction of the 3D surface from the aerial stereo pair "Pentagon". Left stereo image (left), right stereo image (middle), range image of the obtained x-parallax map (right)

"Plain terrain" and "Pentagon" and the reconstructed DEMs as the range images. The range image demonstrates the x-parallax map by means of a greyscale coding of the parallax values—from white for the nearest surface points to black for the most distant ones. The x-parallax map is reduced to the left stereo image (that is, to the raster $R1$). In distinction to the previously proposed in [4–6] symmetric DP algorithms that correspond to the algorithm in Eq. (18) with the weight $w = 1$ the regularized algorithms are far less sensitive to the chosen values of the reconstruction parameters such as e_{min} or d_0 within the above-mentioned range of the weights $0.2 < w < 0.8$.

On the whole, the obtained range images correspond to the visual depth perception of these scenes. However, note that there are several errors in Fig. 5 concentrated mainly along the upper right edge of the Pentagon building. These errors are caused, most likely, by light strip-like details along this edge in the right stereo image. These strips, as Fig. 5 suggests, have no correspondence in the left image and are grouped in the reconstructed profiles with the very similar neighboring strip-like details of the building roof. These details are in both the images.

The errors illustrate the main drawbacks of the simplified symmetric binocular stereo used in the experiments:

(i) it does not take account of y-interactions between the signals because of the profile-by-profile reconstruction mode,
(ii) the underlying difference radiometric model is not suitable to describe stereo images of the surface having parts with metallic luster that can account for the above light strips, and
(iii) the regularizing heuristics cannot overcome the errors caused by the partially occluded and so only monocularly visible surface parts if these latter ones have the coloring that matches closely the neighboring binocularly visible parts (the above strips can result from such an occlusion, also).

Nonetheless, in spite of some local errors, the overall quality of the reconstruction of the observed surfaces is rather good. These and other experimental results show that one can obtain the effective in practice low-level computational static stereo only by integrating the theoretical models and decision rules (for instance, the probabilistic ones) with the various heuristics required to overcome the ill-posedness of the problem. What is more, the heuristic part is to supplement the theoretical part but not substitute for the latter one because of the intricate interactions between the involved data. These interactions cannot be revealed without the in-depth theoretical studies. Thus, the rational tradeoffs between the theoretical foundations and regularizing heuristics are essential to solve the problems of the intensity based static stereo.

Acknowledgements

The author is grateful to Marina Kolesnik for the stereo pair "Plain terrain" and John Weng for the stereo pair "Pentagon". Marina Grigorenko contributed several ideas to select the regularizing heuristics and took part in implementing the program to perform the experiments.

References

[1] Baker, H. H.: Surfaces from mono and stereo images. Photogrammetria *39*, 217–237 (1984).

[2] Barnard, S. T., Fischler, M. A.: Computational stereo. ACM Comput. Surv. *14*, 553–572 (1982).

[3] Hannah, M. J.: Digital stereo image matching techniques. Int. Arch. Photogra. Remote Sensing *27*, 280–293 (1988).

[4] Gimel'farb, G. L.: Symmetric approach to the problem of automating stereoscopic measurements in photogrammetry. Kibernetika *2*, 73–82 (1979) [in Russian; English translation: Cybernetics *8*, 311–322 (1979)].

[5] Gimel'farb, G. L.: Gibbs random fields and compound decisions at the lower level of digital image processing. Pattern Rec. Image Anal. Adv. Math. Theory Appl. USSR *1*, 39–49 (1991).

[6] Gimel'farb, G. L.: Intensity-based computer binocular stereo vision: signal models and algorithms. Int. J. Imaging Syst. Tech. *3*, 189–200 (1991).

[7] Gimel'farb, G. L.: Regularization of low-level binocular stereo vision considering surface smoothness and dissimilarity of superimposed stereo images. In: Proc. of the 2nd Int. Workshop on Visual Form (Capri, Italy, May 30–June 2, 1994), pp. 231–240. Singapore: World Scientific 1994.

[8] Kyreitov, V. R.: Inverse problems of photometry. Novosibirsk: Computing Center (Acad. Sci. USSR, Siberian Branch), 1983. [in Russian]

[9] Marroquin, J., Mitter, S., Poggio, T.: Probabilistic solution of ill-posed problems in computational vision. J. Amer. Stat. Assoc. *82*, 76–89 (1987).

[10] Poggio, T., Torre, V., Koch, C.: Computational vision and regularization theory. Nature *317*, 314–319 (1985).

[11] Schlesinger, M. I.: Mathematical tools for image processing. Kiev: Naukova Dumka, 1989. [in Russian]

[12] Yuille, A. L., Geiger, D., Bulthoff, H.: Stereo integration, mean field theory and psychophysics. In: Lecture Notes in Computer Science 427: Proc. of the First European Conf. on Computer Vision (Antibes, France, April 1990), pp. 73–82. Berlin, Heidelberg, New York, Tokyo: Springer 1990.

Dr. G. L. Gimel'farb
Computer and Automation Research Institute
Budapest
Hungary
e-mail: gimel@miws.ipan.sztaki.hu
(on leave from V. M. Glushkov Institute of
Cybernetics
Kiev
Ukraine)

Computing Suppl. 11, 73–98 (1996)

Surface from Motion—without and with Calibration

R. Klette, Berlin

Abstract

Surface from Motion—without and with Calibration. For uncalibrated and calibrated imaging situations, results on "surface from motion" are given in a systematic order, also described by an abbreviating rule notation. A few new theoretical results for surface from motion are included. Experimental evaluations of reconstruction steps are sketched for some case studies (as optical flow computation, integrative approach to shape from motion, or surface from motion using calibration).

Key words: Surface reconstruction, shape from motion, camera calibration, controlled motion, rotating disc.

1. Introduction

Surface reconstruction via visual sensing is of practical relevance for surface inspection, improved efficiency of input into graphical systems as for modeling objects or animation, or just for generating 3-D views, e.g. for calculating auto-stereograms. Surface reconstruction may be a step within complex solutions for modeling and performing simulation experiments with 3-D objects. Surface reconstruction as the main goal of a vision system may be seen in contrast to active vision [2], but as a subtask it may be also integrated into complex purposive vision systems.

1.1. Informal Task Description

The goal consists in generating a (partial) surface representation as by triangulation or based on a voxel model. Two cases may be considered: *absolute size surface reconstruction,* i.e. in exact scale, or *relative size surface reconstruction* with true distance relations between surface points up to a scale factor. Such (partial) surface representations may be derived from an absolute or dense depth map of surface points. Without camera calibration and/or a-priori knowledge about the object motion in scene space, the "classical approach" consists of two steps: at first a dense shape map is calculated from image features, and then depth is calculated from shape. Here, "shape" is defined to be a set of gradient values of the object surface, cf. [7]. Using calibrated imaging or a-priori knowledge about the type of object motion, depth maps may be directly calculated from image features without going via shape.

In this paper, for uncalibrated and calibrated imaging situations, known results on "surface from motion", cf. [1, 9] are given in a systematic order, also described by the abbreviating rule notation as proposed in [10, 11]. Own experimental evaluations and new theoretical results are included in this review.

The *general problem description* is as follows: Images are taken by a single static camera. Rigid opaque objects are projected into the image plane assuming a certain geometric projection and a certain camera model. These objects may move in 3-D space, and during motion a sequence of images is taken at constant time intervals. As goal, the (visible) object surface has to be reconstructed in 3-D space (absolute size or relative size reconstruction).

For the uncalibrated situation, a pinhole camera with ideal central or parallel (i.e. focal length to infinity) projection is assumed. For the calibrated situation, also some lens distortions are assumed. Here, for calibration only radial lens distortion will be assumed. Device dependent coordinates after digitizing analog image signals have to be considered also for the calibrated situation.

1.2. Formal Projection Model

The following (left handed) coordinate systems will be used to model the relations between scene space objects and projected images, see Fig. 1.1:

(X_w, Y_w, Z_w) denote the 3-D coordinates of object surface points P in the *world coordinate system WCS*,

(X_c, Y_c, Z_c) denote the 3-D coordinates of P in the *camera coordinate system CCS*,

B is the distance of image plane to projection center (*focal length*),

(x_u, y_u) are *undisturbed* image coordinates of (X_c, Y_c, Z_c) assuming an ideal pinhole camera,

(x_d, y_d) are *disturbed* image coordinates, differing from (x_u, y_u) by radial lens distortion, and

(x_f, y_f) are *device-dependent* coordinates of (x_d, y_d) in the digitized image (not illustrated in Fig. 1.1).

The Z-axis Z_c of the camera coordinate system coincides with the optical axis, and is pointing into scene space. For discussions of *controlled object motion* (i.e. availability of a-priori knowledge about this motion), the Z-axis Z_w of the world coordinate system coincides with the rotation axis of a rotating disc.

In the uncalibrated case, only relations between camera coordinates and undisturbed image coordinates will be discussed. Here, world and camera coordinates are identified. For simplification, (x, y) and (X, Y, Z) are used in this case, without indices u and w or c.

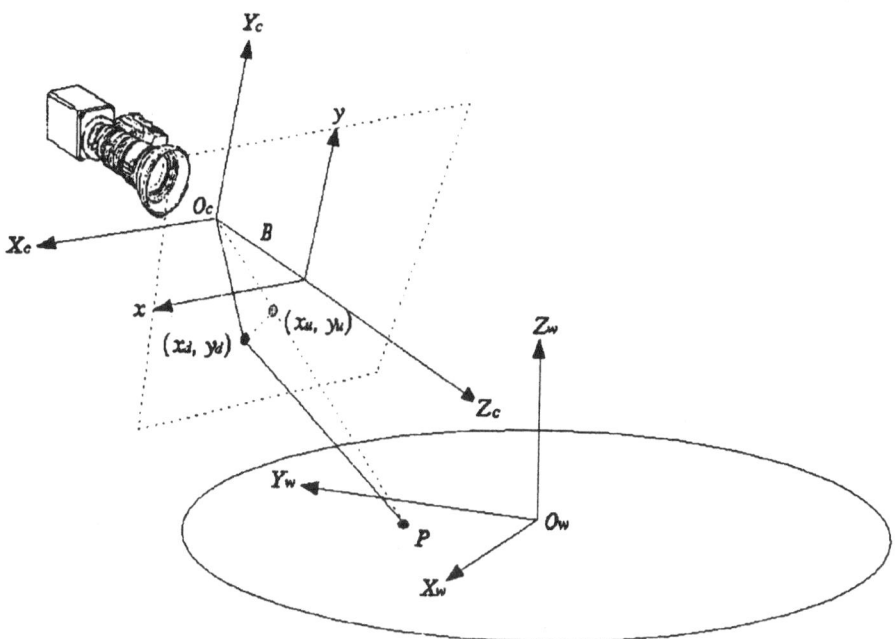

Figure 1.1. Camera geometry with perspective projection: world coordinates $X_w Y_w Z_w$, camera coordinates $X_c Y_c Z_c$, and ideal projected coordinates $x_u y_u$, and radial lens distortion: distorted image coordinates $x_d y_d$. Here, the circular area may be interpreted as an rotating disc, which is used for discussions of controlled motion in Sections 3 and 4

All coordinates and parameters will be measured at the same scale, e.g. μm. The only exception are discrete coordinates (x_f, y_f) for the digitized image which are given in (sub-)pixels.

The 3-D camera coordinates (X_c, Y_c, Z_c) are transformed in ideal, undisturbed image coordinates (x_u, y_u) by perspective projection. According to the pinhole camera model it holds

$$x_u = \frac{B \cdot X_c}{Z_c} \quad \text{and} \quad y_u = \frac{B \cdot Y_c}{Z_c} \qquad (1.1)$$

in the *camera centered coordinate system (ccc)* as shown in Fig. 1.1 where the focal point of the camera, i.e. the projection center, coincides with the origin of the *CCS*. In the *image centered coordinate system (icc)* i.e. the focal point is assumed at $(0, 0, -B)$ in the *CCS*; it holds

$$x_u = \frac{B \cdot X_c}{Z_c + B} \quad \text{and} \quad y_u = \frac{B \cdot Y_c}{Z_c + B}. \qquad (1.2)$$

For B to infinity, parallel projection follows as

$$x_u = X_c \quad \text{and} \quad y_u = Y_c. \qquad (1.3)$$

1.3. Surface Description and Object Motion

An object surface point $P = (X_w, Y_w, Z_w)$ may be in arbitrary position in 3-D space, satisfying $Z_w > B$. Assuming polyhedral approximations, object surfaces are given by planar surface patches. A surface patch is assumed in a plane

$$Z_w = pX_w + qY_w + d. \tag{1.4}$$

The *shape* of the visible surface is defined by the surface normals $n(X_w, Y_w, Z_w)$ or the surface gradients (p, q) of visible surface points (X_w, Y_w, Z_w). The *visible surface* is identified with a (unique) function $Z_w(X_w, Y_w)$, also called the *depth*. For the gradient or shape it holds that

$$p = \frac{\partial Z_w}{\partial X_w} \quad \text{and} \quad q = \frac{\partial Z_w}{\partial Y_w}. \tag{1.5}$$

It is assumed that the surface normals $n(X_w, Y_w, Z_w)$ are always pointing toward the pictorial plane, i.e.

$$n(X_w, Y_w, Z_w) = (p, q, -1). \tag{1.6}$$

By a 3-D motion of a scene object, the motion of one specific surface patch (1.4) may be specified by six parameters $\omega_1, \omega_2, \omega_3, t_1, t_2, t_3$, defining rotational speed $(\omega_1, \omega_2, \omega_3)$ around $(0, 0, d)$ and a subsequent translation (t_1, t_2, t_3),

$$\begin{bmatrix} \dot{X}_w \\ \dot{Y}_w \\ \dot{Z}_w \end{bmatrix} = \begin{bmatrix} t_1 \\ t_2 \\ t_3 \end{bmatrix} + \begin{bmatrix} \omega_1 \\ \omega_2 \\ \omega_3 \end{bmatrix} \times \begin{bmatrix} X_w \\ Y_w \\ Z_w - d \end{bmatrix}. \tag{1.7}$$

Depending upon the projection equation (1.1), (1.2) or (1.3), this surface motion in 3-D space between time t to time $t + 1$ results into a 2-D displacement of ideal image point (x_u, y_u) at time t to point $(x_u + u(x_u, y_u), y_u + v(x_u, y_u))$ at time $t + 1$. In general it holds [9]

$$\begin{aligned} u(x_u, y_u) &= u_0 + A_1 x_u + A_2 y_u + (C_1 x_u + C_2 y_u) x_u \\ v(x_u, y_u) &= v_0 + B_1 x_u + B_2 y_u + (C_1 x_u + C_2 y_u) y_u. \end{aligned} \tag{1.8}$$

For each planar surface patch, these eight parameters $u_0, v_0, A_1, A_2, B_1, B_2, C_1, C_2$ are the *flow parameters*. They may be calculated by LSE minimization for (all) ideal image points of the projection of the surface patch if (!) such a segmentation is available in the image.

By *optical flow* or *local displacement* $u(x_u, y_u) = (u(x_u, y_u), v(x_u, y_u))$, for point (x_u, y_u) at time t the next image point position at time $t + 1$ is specified. All vectors u for (nearly) all image points define the *dense optical flow field* or *dense local displacement field*.

1.4. Subtask Specification

Because of the high complexity of tasks in Computer Vision, even the solutions (algorithms, methods etc.) to basic steps are quite complex normally. These solutions should be stated as precise as possible, including the model assumptions or restrictions of the vision environment which are essential for deriving a solution. Unfortunately, the specification of such assumptions is often not explicitly available in the literature. A formal specification approach by *qualitative rules of reasoning,*

— specifying the assumptions about scenes and the image generation process,
— giving the list of all features which are used to calculate a certain constraint
— for a feature which was the goal of the analyzing process,

was proposed in [10, 11]. These units may be identified with modules in the sense of David Marr [14], but also just with single steps in vision programs, which have to be implemented for active, animate or purposive vision [2], too.

For example, in approaches "shape from *xxx*", *xxx* stands for *shading, shadow, texture, local displacement, line drawing* etc. Going into detail, in general it will be clear that "shape from *xxx*, where Lambertian reflectance, constant albedo, parallel projection, point light source etc. is assumed" characterizes a method reported elsewhere.

For specifying the model assumptions, acronyms will be used for brevity. For the *scene space*, typical assumptions are, e.g., polyhedral objects (*PH*), rigid, opaque bodies (*RO*); 3-D rotation and 3-D translation as 3-D motion (*RTR*), Lambertian reflectance (*LR*), Lambertian reflectance with constant albedo (*LRC*), or Horn-Schunck constraint assumptions (*HS*).

For the *optical projection*, typical assumptions are, e.g., pinhole camera (*pin*), Z-axis of *CCS* identical with optical axis, and pointing into scene space (*pos*), Z-axis of *CCS* identical with optical axis, but pointing toward pictorial plane (*neg*), central projection (*cp*) with camera-centered coordinate system (*ccc*), or with image-centered coordinate system (*icc*), parallel projection (*pp*), single point light source (*ps*), or single point light source in constant direction to surfaces (*psc*). Mentioning cases as *pos* or *neg*, and *ccc* or *icc* is of formal importance.

Rules are verified by *combining theorems* which may also be used to design *derivational algorithms.* Rule, theorem, and algorithm constitute a *derivational unit.* A rule may be formulated following the general syntax pattern

$$[SCENE\ SPACE;\ transform]:\ premise \rightarrow conclusion.$$

In *premise* and *conclusion*, certain 2-D or 3-D features of scene objects or scene space may be listed. A combining theorem gives some mathematical dependencies between different features, it "combines these features in some sense", for "continuing visual tasks". A derivational algorithm applies such a theorem to some extent for specifying a feature listed in *conclusion.* Two examples: For the (positive) rule

$$[RO, PH, RTR; cp]:\ (3\text{-}D\ rotation,\ 3\text{-}D\ translation,\ local\ displacement) \rightarrow shape$$
$$(1.9)$$

Propositions 3.1 (*ccc*) and 7.1 (*icc*) in [9] may be cited as combining theorems. In this rule, the features in the *premise* belong to the same feature category *MOTION*. There may be different derivational algorithms based on such a combining theorem. The (negative) rule

$$\text{NOT}([LR; ps, pp]: \textit{shading in one image} \rightarrow \textit{lighting direction}) \qquad (1.10)$$

is verified by Theorem 1, Section 3.4; in [1] as combining theorem.

1.5. Contents of the Paper

The paper is structured as follows: For surface from motion, the computation of local displacement fields is of essential importance. A specific approach for evaluation of optical flow algorithms is reported in Section 2. Some results for quantitative evaluations of optical flow computations are given. For central projection, knowledge about the type of motion may be of practical importance for constraining correspondence based motion detection.

In Section 3, the uncalibrated case will be dealt with. Here, at first parallel projection, and then central projection will be considered. For both assumptions, optical flow results are treated as inputs for shape from motion. Based on dense shape maps, relative depth may be calculated, leading to a (partial) 3-D surface representation. For parallel projection and central projection, also two integrative approaches, i.e. several feature categories in *premise*, will be sketched. For such an integrative approach, the importance of studying different derivational algorithms for one combining theorem will be illustrated.

In Section 4, for the calibrated case a specific calibration method was selected and modified at one point. It will be shown how calibration results may be applied for surface reconstruction if object rotation is assumed. For this new method, some results on accuracy and robustness evaluation are added.

Finally, in Section 5 the conclusions are given. Surface reconstruction is characterized by high accuracy demands. By integrating surface from motion and surface from photometric properties, it should be possible to satisfy these practical demands.

2. Calculation of Optical Flow Fields

Computation of high-accuracy dense motion vector fields is the most critical point of surface from motion, cf. [4–6] for evaluations of optical flow algorithms. Approaches for solving this task may be classified as point-based differential methods, region based matching methods, contour-based methods, or energy-based methods.

In the papers [5, 6] a quantitative evaluation of several point-based differential methods is given where it is assumed that spatial and time gradients of the image irradiance function may be calculated. The point-based differential methods were selected because complete and dense motion fields are required for surface from motion approaches. The experiment specifications were as follows:

Input images and ground truth: Synthetic images were generated for simulating translations and rotations. In experiment (A), motion of textured (autoregressive pseudo-random patterns) planes were studied. In experiment (B), planar circles were moving on a differently textured plane in 3-D space. Translation and/or rotation parameters were available as ground truth.
Error measure: For the ground truth $\mathbf{u}^* = (u^*, v^*)$ of the motion model and the calculated image velocity $\mathbf{u} = (u, v)$, the *normalized sum of all relative errors* (SRE) between $\mathbf{u}^*$ and $\mathbf{u}$ was used as error measure.

For experiment (A), some typical results are illustrated in Fig. 2.1. The different methods behave differently on generated textures. Surprisingly (?), the original Horn/Schunck-method was quite tolerant to the different textures. The Nagel-method is of theoretical interest only (it uses the unknown 3-D positions of surface points), and did not show any essential improvement in comparison to the Horn/ Schunck-method. In general, all these differential methods fail if there is not sufficient "image value structure" in the texture. As extreme case, surfaces with about constant coloring will lead to no usable result at all (as anyone will expect).

For experiment (B), some typical results are shown in Fig. 2.2. Here, for series of differently textured circles moving on textured planar backgrounds (inverse motions were used for generating "simple" motion boundaries), the errors did increase in comparison to experiment (A). Even after improving the initialization of the iterative methods by using a result of a non-iterative pseudo-inverse method for initialization, the SRE error was about 5% at best, where the Nagel/Enkelmann-

lows:

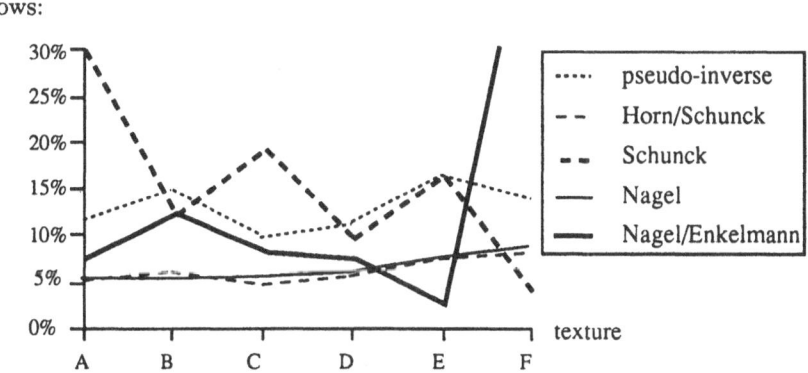

Figure 2.1. SRE error for experiment (A) of six differently textured planes where a small (constant) translation of these planes was simulated

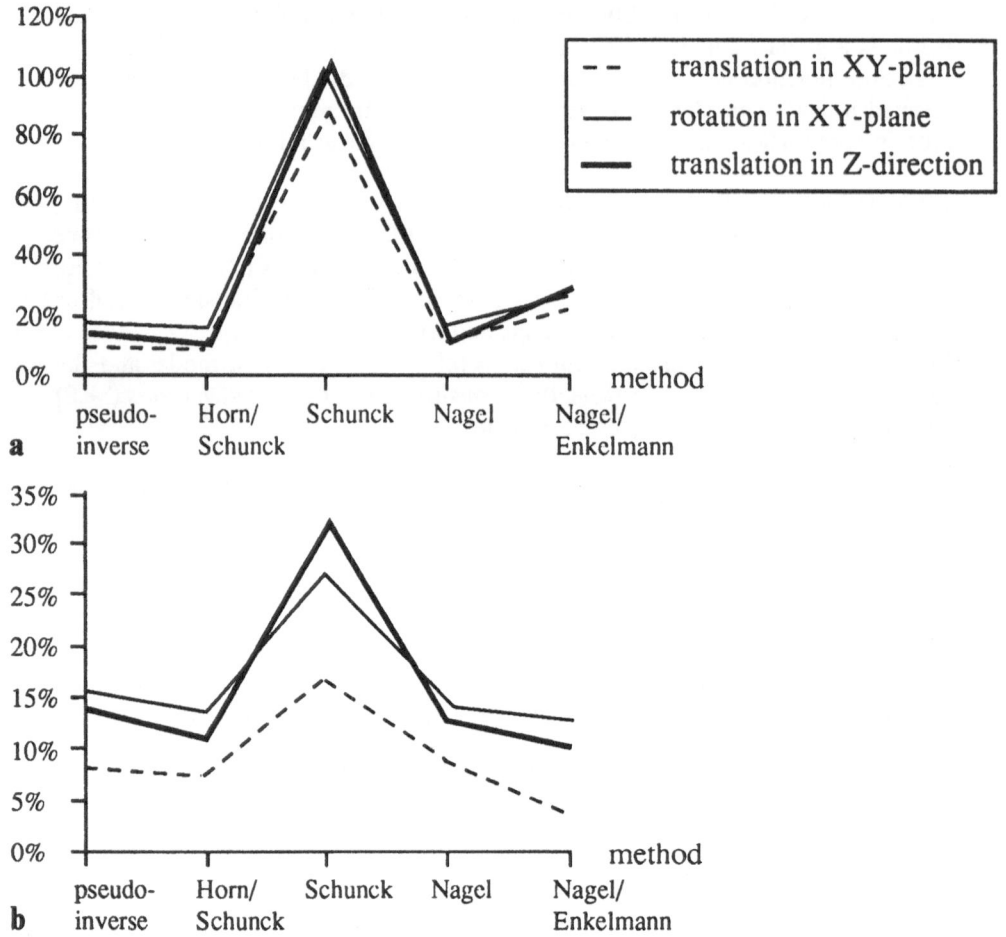

Figure 2.2. Averaged SRE errors for experiment (B) for moving planar textured circles and inverse motions of the textured planar backgrounds. **a** Start values $(0, 0)^T$ for the iterative methods. **b** Results of the pseudo-inverse method were used as start values for the iterative methods

method [18] did behave best for this two-component motion. Note that these homogeneous textured circles on a homogeneous textured background still represent the "simplest input" for such a motion detection algorithm. The iterative Schunck-method was even worse then the used non-iterative pseudo-inverse method!

Later on, also further motion detection algorithms, not listed in Fig. 2.1 or Fig. 2.2, were evaluated [16] using the source code of [4]. Because real objects often do not have "nicely textured surfaces", the experimental results were even worse! Thus, for the following results on shape from motion, erroneous input data have to be taken in mind if images of real moving objects have to be analyzed.

Based on a-priori knowledge about object motion, correspondence based motion detection may be constrained for obtaining improved results. Assume that objects are placed on a rotating disc, cp. Fig. 1.1, i.e. radius of disc and a fixed rotation axis are given (for calculating the axis, camera calibration may be used). For this case of rotational motion [16], for correspondence calculation the epipolar constraint of static stereo may be adopted. In this specific (partially calibrated) case, optical flow vectors connecting corresponding image points may be used to calculate depth without going via shape.

3. Reconstruction without Calibration

In the uncalibrated case, shape from motion is proposed in literature as one of the general computer vision modules [7], and even some books in computer vision are focusing on that issue, e.g. [9, 15]. At first, for a time sequence of projections, motion vectors have to be computed. Then, based on these vectors, certain shape values of the objects may be determined. Finally, from shape values some depth information as to be derived leading to a certain 3-D representation of the projected objects.

3.1. Parallel Projection

In a specific vision task, parallel or central projection may be assumed as relevant model. As already mentioned in Section 2, the "parallel projection method" by Horn and Schunck, and the "central projection method" by Nagel did lead to about the same optical flow fields. In subsequent steps of shape from motion and depth from shape, the *pp* model may simplify computations.

3.1.1. Shape from Motion: the General Case *pp*

In this Subsection, three "famous" results are scheduled which seem to be of theoretical interest only for surface reconstruction remembering the inaccuracy of optical flow fields as briefly discussed above.

At first, assuming parallel projection, the flow parameters of (1.8) are given by the following Theorem. Here, six non-trivial equations are available to constrain seven object and motion parameters $p, q, t_1, t_2, \omega_1, \omega_2, \omega_3$.

Theorem 3.1 [9]. *In case of* $[RO, PH, RTR; pp]$ *it holds*

$$u_0 = t_1 \qquad\qquad v_0 = t_2$$
$$A_1 = p\omega_2 \qquad\qquad A_2 = q\omega_2 - \omega_3$$
$$B_1 = -p\omega_1 + \omega_3 \qquad\qquad B_2 = -q\omega_1$$
$$C_1 = 0 \qquad\qquad C_2 = 0.$$

This combination theorem for *3-D* motion, local displacement and shape, taking into account all assumptions as used in the proof, leads, e.g., to the rule

$[RO, PH, RTR; pp]$: (*3-D rotation, 3-D translation, local displacement*) $\rightarrow$ *shape*.

(3.1)

But algorithms based on Theorem 3.1 do not lead to unique shape values if no further techniques as e.g. regularization are applied. What is more, he computation of flow parameters from the flow field is numerically complicated, optical flow values at several image points within the projection of the same planar surface patch must be available for using optimization or regularization. And: Segmentations of the images into projections of planar surface patches have to be available.

Let u_x, u_y, v_x, v_y be the following displacement differences (this is not a "naive" gradient approximation) in case *neg*,

$$u_x = u(x + 1, y) - u(x, y), \qquad v_x = v(x + 1, y) - v(x, y),$$
$$u_y = u(x, y + 1) - u(x, y), \qquad v_y = v(x, y + 1) - v(x, y).$$

(3.2)

In the case *pos*, these displacement differences are defined by

$$u_x = u(x + 1, y) - u(x, y), \qquad v_x = v(x + 1, y) - v(x, y),$$
$$u_y = u(x, y - 1) - u(x, y), \qquad v_y = v(x, y - 1) - v(x, y).$$

(3.3)

ensuring the "same orientation of the differences with respect to the Z-axis". By using the displacement differences for the case *pos*, from Theorem 3.1 it follows

Corollary 3.2. *For $\omega_1, \omega_2 \neq 0$, the shape parameters may be uniquely calculated by*

$$p(x, y) = \frac{\omega_3 - v_x(x, y)}{\omega_1} = \frac{u_x(x, y)}{\omega_2}$$

and

$$q(x, y) = \frac{u_y(x, y) + \omega_3}{\omega_2} = \frac{-v_y(x, y)}{\omega_1}.$$

Furthermore, it holds

$$\frac{\omega_1}{\omega_2} = \frac{v_y}{u_x} \cdot \frac{\dfrac{v_y - u_y + \sqrt{(v_x + u_y)^2 - 4u_x v_y}}{2} - v_x}{v_y}.$$

This is a combining theorem for a rule

$[RO, PH, RTR; pp]$: (*3-D rotation, local displacement*) $\rightarrow$ *shape*. (3.4)

Corollary 3.2 coincides (in principle) with the first two expressions of Theorem 3.b in [1], but in the second expression a negative sign was obtained. Here, the difficulty of calculating flow parameters for applying Theorem 3.1 is "replaced" by the difficulty of obtaining correct (subpixel accuracy) differences of neighboring motion vectors.

Derivational algorithms based on Corollary 3.2 will be even more sensitive to incorrect motion vector results as algorithms based on motion vectors itself. This sensitivity is also relevant for

Theorem 3.3 [1]. *In case of* $[RO; pp]$ *it holds*

$$a \cdot p^2 + b \cdot q^2 - 2c \cdot pq + d = 0$$

where

$$a = u_y^2 + v_y^2 + 2v_y, \qquad\qquad b = u_x^2 + v_x^2 + 2u_x,$$

$$c = u_y + u_y \cdot u_x + v_x + v_x \cdot v_y \quad and \quad d = c^2 - ab.$$

In comparison to (3.1) and (3.4), this combining theorem is proving the most general rule,

$$[RO; pp]: local\ displacement \rightarrow shape. \tag{3.5}$$

The polynomial of second order may be solved, e.g., algebraically, numerically, or by optimization techniques, i.e. for this combining theorem, very different derivational algorithms, for derivation of shape from motion, may be developed. This will be illustrated more in detail in 2.3.1.

3.1.2. Shape from Motion, Lighting Direction and Shading

Integrative approaches are characterized by several feature categories in the *premise* of a derivation rule. By Theorem 2, Section 3.5, in [1], an interesting integrative combining theorem is given (which has to be slightly corrected). For two intensity images f_1 and f_2 from a series of images, the constant direction

$$\mathbf{s} = (s_1, s_2, s_3)$$

from scene space objects to a point light source, the local displacement $u(x, y)$ and $v(x, y)$, and from the ratio

$$r(x, y) = \frac{f_2(x + u(x, y), y + v(x, y))}{f_1(x, y)}, \quad \text{for } f_1(x, y) \neq 0, \tag{3.6}$$

of intensity values of "moving image points", a constraint for shape (p, q) was derived. As model assumptions, Lambertian reflection

$$f(x, y) = \rho \cdot \mathbf{n}(x, y) \cdot s \tag{3.7}$$

with constant albedo ρ is also used.

Assume case *neg*. Then, for the displacement differences u_x, u_y, v_x, v_y the following abbreviations will be used:

$$A = (u_x + 1)(v_y + 1) - v_x u_y, \qquad B = u_y s_2 - s_1(v_y + 1),$$

$$C = v_x s_1 - s_2(u_x + 1), \qquad D = u_y^2 + v_y(v_y + 2),$$

$$E = v_x^2 + u_x(u_x + 2) \qquad and \quad F = 2u_y(u_x + 1) + 2v_x(v_y + 1). \tag{3.8}$$

Theorem 3.4 (cf. [1]). *In case neg it holds*

$$ap^2 + bq^2 + cpq + dp + eq + f = 0$$

with

$$a = r^2 s_1^2 - B^2, \qquad\qquad b = r^2 s_2^2 - C^2,$$

$$c = 2[r^2 s_1 s_2 - BC], \qquad d = 2rs_1 s_3(r - A),$$

$$e = 2rs_2 s_3(r - A) \qquad and \quad f = s_3^2[A(A - 2r) + r^2] + C^2 D + BCF + B^2 E.$$

Again, this combining theorem is based on displacement differences making its practical use for shape reconstruction highly questionable (originally, it was suggested for estimating the direction to the point light source). But, even for the case that absolute correct optical flow vectors are given, derivational algorithms for calculating shape, based on this Theorem 3.4, will produce some errors by its own. In [12], three algorithmic realizations for the rule

$$[RO, LRC; psc, pp]: (shading, local\ displacement, lighting\ direction) \rightarrow shape, \quad (3.9)$$

proved by Theorem 3.4, were reported, and the resulting errors of calculated shape were discussed. For details of the algorithms, see this cited conference paper.

By testing this approach on synthetic images, some detailed studies were possible. Because of the small values of displacement differences, roundings in the used algebraic expressions (as changes between *neg* and *pos*) have dramatic impact on the obtained practical results. The quality of the computable shape parameters depends upon the complexity of the 3-D motion, where complex motion leads to better results. If only translation is assumed, then the results are very poor. This follows also by Theorem 2.1: the additive constant disappears in the difference functions.

The second-order polynomial in p and q in Theorem 2.4 allows four different algebraic solutions (p, q) in general. Thus, by three different polynomials (constraints) a unique solution is possible, and more than three constraints lead to some optimization. A general way for solving this polynomial is as follows:

(i) Starting with a first constraint $a_1 p^2 + b_1 q^2 + c_1 pq + d_1 p + e_1 q + f_1 = 0$ for images f_0 and f_1, two solutions $p = p(q)$ may be calculated.
(ii) By insertion of these solutions into a second constraint $a_2 p^2 + b_2 q^2 + c_2 pq + d_2 p + e_2 q + f_2 = 0$ for images f_0 and f_2, at the same point locations (x, y), a fourth order polynomial in q follows.
(iii) For the algorithmic solution of this fourth order polynomial, different methods may be applied leading to different *derivational algorithms* for deriving shape from shading, local displacement, and lighting direction:

Algorithm 1 (algebraic method): There are four (complex) solutions of this polynomial. The computed four values of q are used to calculate eight related values of p. A unique final solution may be computed by implementing evidence rules.

Algorithm 2 (numeric method): The fourth order polynomial in q may be solved by a numeric iterative procedure following [17].

Algorithm 3 (LSE optimization): For a sequence of images $f_0, f_1, f_2, \ldots$, image f_0 is taken as reference where for points (x, y) in image f_0 the gradients have to be computed. The difference of

$$error_i(p, q) = ap^2 + bq^2 + cpq + dp + eq + f \qquad (3.10)$$

to zero evaluates a gradient (p, q), and during processing of the image sequence, the arithmetic mean value of squares of these errors $error_i(p, q)$ will be minimized.

The experiment specifications were as follows:

Input images and ground truth: Several synthetic objects in motion were used as input. The synthetic object of a Lambertian sphere allows a good comparison for all directions of surface normals in visible surface points. Object motion was known as ground truth. The position of the point light source, i.e. vector s, was assumed to be identical for each triplet of runs of the three different algorithms. This position of the light source has some influence to the results in that sense that surface normals with directions orthogonal to the direction of vector s are difficult to compute (this follows directly from Theorem 3.4).

Error measure: For interactive evaluation, the resulting needle maps of difference vectors (!) between surface normals and computed normals (arithmetic average of errors) were graphically represented on a sphere. If the object itself was a sphere, a dense picture of these "error vectors" could be obtained.

In Fig. 3.1 for the (best) case of the LSE-optimization, the two diagrams have the following interpretation. In the upper diagram, the differences of computed normals to the ideal normals are illustrated for the p-direction, i.e. on the left the errors for normals in directions $-90° \ldots 0°$, and on the right the errors for normals in directions $0° \ldots 90°$ are shown. The p-direction means that the left half of the diagram illustrates the left semisphere, and the right half the right semisphere. The lower diagram corresponds to the q-direction, i.e. on the left the lower semisphere, and on the right the upper semisphere is illustrated. The errors are shown as values between $0°$ and $90°$.

As shown in Fig. 3.1, the errors increase if surface plane and picture plane are close to being nearly orthogonal. The direction s to the point light source has essential influence on the computed normals. The critical situation is given if the direction s coincides with the optical axis, i.e. the Z-axis, because in this case $a = b = \cdots = f = 0$ follows for Theorem 3.4. In case of the sphere, see Fig. 3.1, the normals on a diagonal which is orthogonal to direction s, are hard to recognize because coefficients d, e and f of the constraint are close to zero. Also, the 3-D motion of the objects has essential impact. There should be rotations with respect to all three coordinate axis. Normals orthogonal to the motion direction may be recognized very stable.

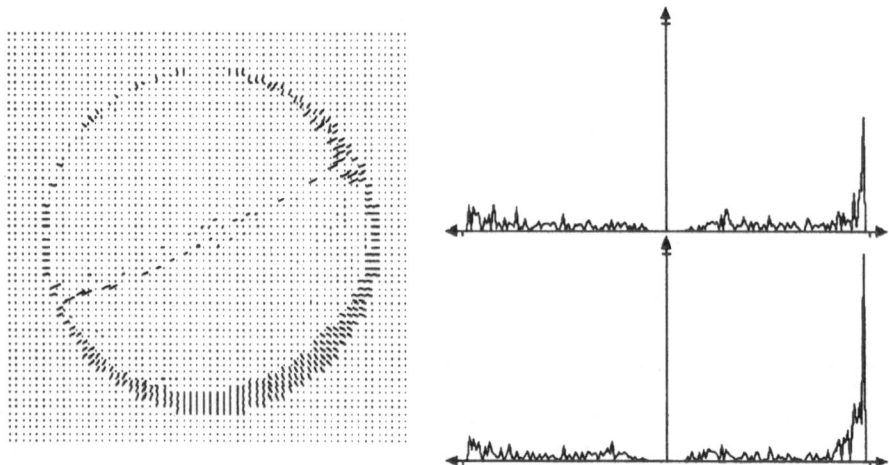

Figure 3.1. Needle map of difference vectors (!) and differences between computed and actual normals in case of using the LSE optimization

By comparing these three derivational algorithms, the algebraic method is computationally fast, very instable, and imaginary numbers appear and have to be approximated. By this algorithm, only normals $\mathbf{n} = (p, q, -1)$ may be recognized which form an angle less then 90° with the direction s to the point light source.

The numeric method is more stable with respect to the number of solutions (different picture points (x, y)) and their quality. If surface plane and picture plane are close to being nearly orthogonal, erroneous results may be avoided by a threshold. Imaginary terms during iteration are taken as they are, and not approximated. Typically, a very fast convergence was considered. An additional treatment of the results by a Newton iteration did not lead to essential improvements.

The LSE-optimization has given in all experiments the best results. By specifying the resolution during minimization, the quality of the results may be selected according to possible computing time limitations. Even in the case of a coarse resolution, say 10°, the computed gradients are quite accurate. Drawbacks of this algorithm are large time complexity and missing solutions for gradient directions close to 90°.

3.1.3. Depth from Shape

On the way from images to surfaces, the next problem is connected with going from shape (gradients) to depth, i.e. exact surface point positions in 3-D space.

For pp, a dense shape map may be used to obtain depth estimates up to a constant summand. Because of $(x, y) = (X, Y)$, it follows

$$(p, q) = \left(\frac{\partial Z}{\partial X}, \frac{\partial Z}{\partial Y} \right) = \left(\frac{\partial Z}{\partial x}, \frac{\partial Z}{\partial y} \right). \tag{3.11}$$

Thus, using only the linear terms in the Taylor expansion

$$Z(x + \delta x, y + \delta y) - Z(x, y) = \frac{\partial Z}{\partial x} \cdot \delta x + \frac{\partial Z}{\partial y} \cdot \delta y + \cdots \qquad (3.12)$$

the given values of p, q and δx, δy are sufficient for calculating a relative depth map. This is already a combining theorem for a rule as

$$[pp]: shape \to depth. \qquad (3.13)$$

Theoretically, in general by a few absolute depth values the relative map should be transformable into an absolute depth map.

A straightforward approach for a derivation algorithm is as follows: Based on shape values, depth will be propagated in row and column direction. Then, from both processes the average will be used as final value.

Such an algorithm should be refined to react also on local gradient distributions. But e.g. for a gradient image of stairs (with incomplete shape information) it will always remain to be impossible to transform this into a perfect surface reconstruction.

3.2. Central Projection

For models ccc or icc there are specific (formal) reasons when to use which model [9]. In the general case of cp, no knowledge about the type of motion, or even motion parameters is available. As in the general case of pp, at first local displacement fields have to be computed, then shape from motion, followed by depth from shape, and surface from depth.

3.2.1. Shape from Motion: The General Case cp

In this Section, two basic results for shape from motion are used for illustrating the situation in this general case. Assuming central projection and icc, the flow parameters of (1.8) are given as follows:

Theorem 3.5 [9]. *In case of* $[RO, PH, RTR; icc]$ *it holds*

$$u_0 = \frac{Bt_1}{B + d}, \qquad v_0 = \frac{Bt_2}{B + d},$$

$$A_1 = p\omega_2 - \frac{pt_1 + t_3}{B + d}, \qquad A_2 = q\omega_2 - \omega_3 - \frac{qt_1}{B + d},$$

$$B_1 = -p\omega_1 + \omega_3 - \frac{pt_2}{B + d}, \qquad B_2 = -q\omega_1 - \frac{qt_2 + t_3}{B + d},$$

$$C_1 = \frac{1}{B}\left(\omega_2 + \frac{pt_3}{B + d}\right), \qquad and \quad C_2 = \frac{1}{B}\left(-\omega_1 + \frac{qt_3}{B + d}\right).$$

For an image sequence, at first eight flow parameters u_0, v_0, A_1, A_2, B_1, B_2, C_1, C_2 have to be computed from the optical flow field, and then these eight parameters may be used, following Theorem 3.5, for constraining nine 3-D parameters t_1, t_2, t_3, ω_1, ω_2, ω_3, p, q, and d. By this combining theorem, the rule

$$[RO, PH, RTR; cp]: (3\text{-}D \text{ rotation}, 3\text{-}D \text{ translation}, \text{local displacement}) \rightarrow \text{shape}$$
(3.14)

is proved, i.e. shape is constrained by *3-D rotation, 3-D translation* and *displacement* under the model assumptions listed in the premise. Similarly as Corollary 3.2 follows from Theorem 3.1, here it follows that shape (p, q) as well as even depth d may be calculated by using the following four second order polynomials in p, q and d.

Corollary 3.6 [11]. *Shape and depth are constrained by*

$$B^2(u_y + \omega_3) + B\omega_1 x = q[B^2\omega_2 - Bt_1 + t_3 x] + qdB\omega_2 - d[Bu_y + B\omega_3 + \omega_1 x],$$

$$B^2(v_x - \omega_3) - B\omega_2 y = p[-B^2\omega_1 - Bt_2 + t_3 y] - pdB\omega_1$$
$$- d[Bv_x - B\omega_3 - \omega_2 y],$$

$$B(Bu_x + t_3 - 2\omega_2 x - \omega_2 + \omega_1 y) = p[B^2\omega_2 - Bt_1 + 2t_3 x + t_3] + qt_3 y$$
$$+ d[-Bu_x + 2\omega_2 x + \omega_2 - \omega_1 y] + pdB\omega_2,$$

$$B(Bv_y + t_3 - 2\omega_1 y - \omega_1 + \omega_2 x) = pt_3 x + q[-B^2\omega_1 - Bt_2 + 2t_3 y + t_3]$$
$$+ d[-Bv_y - 2\omega_1 y - \omega_1 + \omega_2 x] - qdB\omega_1.$$

By solving this algebraic equation system in the form

$$A_1 = B_1 q + C_1 qd - D_1 d \qquad A_2 = B_2 p - C_2 pd - D_2 d$$
$$E_1 = F_1 p + G_1 q + H_1 d + C_1 pd \qquad E_2 = G_2 p + F_2 q + H_2 d - C_2 qd,$$
(3.15)

a third-order polynomial for d results. Because of the inaccuracy of computed optical low fields (cf. Section 2), and the numeric problems of solving complex equation systems or non-linear polynomials (cf. 3.1.2), these results seem to be of theoretical interest only. This is also due to the fact that Theorems 3.1 and 3.5, as well as Corollaries 3.2 and 3.6 are formulated for a specific planar surface patch, i.e. the projection of this patch has to be properly segmented in the image.

3.2.2. Shape from Motion and Texture

In the following approach to texture based shape-from-motion analysis as proposed by [1], the model assumptions are considerably general (besides cp, also pp may be assumed, but here it will be discussed for cp). Here it is required, that a "nearly correct" flow field is available, the focal length B must be known, and an estimation procedure for the motion parameters of the visible object surfaces should be available.

Assume surface patch motion as specified in Subsection 1.3. In general, for projections the representations

$$\dot{x} = \mathbf{u}(x, y) = \sum_{i=1}^{6} b_i \cdot u_i(p, q, d, x, y)$$

$$\dot{y} = \mathbf{v}(x, y) = \sum_{i=1}^{6} b_i \cdot v_i(p, q, d, x, y)$$

(3.16)

may be assumed, here for the six motion parameters $b_1 = t_1$, $b_2 = t_2$, $b_3 = t_3$, $b_4 = \omega_1$, $b_5 = \omega_2$, $b_6 = \omega_3$ of a specific surface patch.

Theorem 3.7. *In case of ccc with focal length B it holds that*

$$u_1(p, q, d, x, y) = \frac{B(B - px - qy)}{Bd} \qquad v_1(p, q, d, x, y) = 0$$

$$u_2(p, q, d, x, y) = 0 \qquad v_2(p, q, d, x, y) = \frac{B(B - px - qy)}{Bd}$$

$$u_3(p, q, d, x, y) = \frac{-x(B - px - qy)}{Bd} \qquad v_3(p, q, d, x, y) = \frac{-y(B - px - qy)}{Bd}$$

$$u_4(p, q, d, x, y) = \frac{-xy}{B} \qquad v_4(p, q, d, x, y) = -px - qy - \frac{y^2}{B}$$

$$u_5(p, q, d, x, y) = px + qy + \frac{x^2}{B} \qquad v_5(p, q, d, x, y) = \frac{xy}{B}$$

$$u_6(p, q, d, x, y) = -y \qquad v_6(p, q, d, x, y) = x$$

These equations differ from that given in Section 4.3 of [1]. They proposed to measure texture within a region **B** of an intensity image f by linear features

$$M_i = \int_{\mathbf{B}} \int f(x, y) m_i(x, y) \, dx \, dy.$$

(3.17)

By $m_0(x, y) = 1$, $m_1(x, y) = x$, $m_2(x, y) = y$, $m_3(x, y) = x^2$, $m_4(x, y) = xy$ etc. an example is given (moments of region **B**).

For the changes of these texture features during the motion process it follows that

$$\dot{M}_i = -\int_{\mathbf{B}} \int m_i(x, y) [-(f_x, f_y)(\dot{x}, \dot{y})] \, dx \, dy$$

$$= \sum_{k=1}^{6} b_k h_{ik}(p, q, d, x, y)$$

(3.18)

with

$$h_{ik}(p, q, d, x, y) = -\int_{\mathbf{B}} \int m_i(x, y) [u_i(p, q, d, x, y) f_x$$

$$+ v_k(p, q, d, x, y) f_y] \, dx \, dy.$$

(3.19)

Here, f_x and f_y denote the derivations of image f in x- and y-direction, and the Horn-Schunck constraint was used to replace $\dot{f}$ by $-(f_x, f_y)(\dot{x}, \dot{y})$. Thus, as model assumptions we have to remember this precondition (HS).

The functions h_{ik} may be calculated for the assumed projection, cp. [10]. Then, by the derivations $\dot{M_i}$ of the linear texture features, which may be assumed to be known by low-level image processing, for shape and depth d an equation system will result. Altogether, this approach may be described by the rule

[{*motion*}, *HS*; {*projection*}]:

$$(motion\ parameters,\ \{linear\ texture\}) \rightarrow shape \qquad (3.20)$$

where {*motion*} may be any specific 3-D motion in scene space, {*projection*} a specific projection of scene space into the pictorial plane, and {*lineare texture*} a specific way to measure linear texture features. This illustrates, that the consideration of qualitative rules of reasoning may be also extended to rules containing model variables with defined domains.

In the final theoretical result of this approach (equation systems in 3-D motion parameters, p, q, d, x, y and $\dot{M_i}$'s), the flow field is not contained anymore. It was just used for proving this result. Here, the derivations of the linear features might be critical (calculation for corresponding surface patches in the image sequence). But, still it is suggested to remember this approach for shape and/or depth calculation in future work. For controlled motion, the 3-D parameters may be given or calculated. For polyhedral objects, the correspondence problem of surface patches may be solvable.

3.2.3. Depth from Shape

For cp, depth may be determined by shape up to a multiplicative constant under some assumptions; the concrete formulae, cp [1], are given in the following Theorem 3.8. In the proof of this Theorem it was assumed that $\delta x \cdot \delta Z \approx 0$ and $\delta y \cdot \delta Z \approx 0$, i.e., no abrupt changes of depth abbreviated by SMD ("smooth depth distributions") and that

$$Z(X + \delta X, Y + \delta Y) = Z(X, Y) + p \cdot \delta X + q \cdot \delta Y, \qquad (3.21)$$

i.e. the non-linear term of the Taylor expansion is also assumed to be zero, i.e. locally linear surfaces (LLS). In this situation it holds

Theorem 3.8. *For cp, SMD and LLS it holds that*

$$\frac{Z(X + \delta X, Y + \delta Y)}{Z(X, Y)} \approx 1 + c(X, Y)$$

for

$$c(X, Y) = \frac{p}{B - px - qy} \delta x + \frac{q}{B - px - qy} \delta y.$$

As *constraint* it follows from this that for known shape parameters p and q along a path in the pictorial plane the depth ratio may be calculated, i.e., the depth values are uniquely defined during path following if a start value is given. A complete rule proved by this combining theorem is

$$[SMD, LLS; cp]: shape \rightarrow depth. \tag{3.22}$$

Based on this Theorem, as straightforward algorithmic solution the same approach may be suggested as in Subsection 3.1.3.

4. Reconstruction with Calibration

A good review on calibration techniques is given in [19]. As controlled object motion, objects are placed on a rotating disc in front of a single static camera, cf. Fig. 1.1. Here, reconstruction will be supported by accurate camera calibration which has to be realized as preprocessing. The method by [19, 20] was selected and extended [13].

4.1. Camera Calibration

For camera calibration, internal camera parameters as well as geometric relations between camera coordinates and world coordinates have to be pre-calculated, and these data have essential influence on the accuracy obtainable in surface reconstruction. The selected method may be classified to be a non-linear optimization technique.

The Z-axis Z_w of the world coordinate system coincides with the rotation axis of the rotating disc. This is not of essential importance for calibration, but will simplify the used approach to surface reconstruction.

Four steps are considered for mapping an object surface point (X_w, Y_w, Z_w) onto device dependent coordinates (x_f, y_f):

(1) For the affine transform from world coordinates (X_w, Y_w, Z_w) into camera coordinates (X_c, Y_c, Z_c), a rotation matrix $\mathbf{R}$ and translation vector $\mathbf{T}$ have to be calibrated.
(2) For transforming the 3-D camera coordinates (X_c, Y_c, Z_c) in ideal, undisturbed image coordinates (x_u, y_u) by perspective projection, the focal length B has to be calibrated.
(3) For the calculation of undisturbed image coordinates (x_u, y_u) from real, distorted image coordinates (x_d, y_d) the calibration method as proposed by [19] had to be extended. The equations

$$x_d + D_x = u_x, \qquad y_d + D_y = y \tag{4.1}$$

are based on the following abbreviations (for radial distortion):

$$D_x = x_d \cdot (\kappa_1 r^2 + \kappa_2 r^4), \qquad D_y = y_d \cdot (\kappa_1 r^2 + \kappa_2 r^4)$$
$$\text{and} \quad r = \sqrt{x_d^2 + y_d^2}. \tag{4.2}$$

The *distortion coefficients* κ_1 and κ_2 have to be calibrated. A positive value of κ_1 or κ_2 means that some stretching has to be performed for going from distorted to undisturbed coordinates.

Different lens aberrations or distortions may be characterized by an infinite sequence of coefficients. Here, only the first two coefficients are considered. Otherwise, numeric instability would influence the result. Equations (4.2) may be used for the restoration of images if values of (x_d, y_d) are known. But, for going from undisturbed to distorted coordinates, non-linear equations will result. This direction is not described in [19]. Applying numeric methods for solving non-linear equation systems leads to non-acceptable inefficiency, e.g. if a complete image has to be transformed. For that reason, for distorting ideal image points (x_u, y_u) the following simple approximation was used:

$$x_{di} = \frac{x_u}{1 + \kappa_1 r_{(i-1)}^2 + \kappa_2 r_{(i-1)}^4}, \qquad y_{di} = \frac{y_u}{1 + \kappa_1 r_{(i-1)}^2 + \kappa_2 r_{(i-1)}^4}$$
$$\text{with} \quad r_i = \sqrt{x_{di}^2 + y_{di}^2} \quad \text{for} \quad i \in \{1, \dots, n\}. \tag{4.3}$$

With initial value $r_0 = \sqrt{x_u^2 + y_u^2}$, (4.3) leads to a first approximation of the desired solution (x_d, y_d). By iteration, improved radii r_i may be calculated.[1] This iterative procedure represents an extension of the original *Tsai* method.

(4) For transforming real or distorted image coordinates (x_d, y_d) into device dependent image coordinates (x_f, y_f), several parameters have to be calibrated as described in [19].

The accuracy of this calibration method was evaluated using different calibration objects (i.e. planes with calibration points), cf. [16]. For three calibration planes, each with 20 calibration points ("open cube") the accuracy is already close to the optimum.

4.2. Depth from Correspondence

By using dynamic stereo based on the rotating disc, for corresponding points (or: initial point and its flow vector) in consecutive images depth may be computed directly without going via shape.

Assume that during taking images of the object placed on the rotating disc, projections C_1 and C_2 in the image plane of the *CCS* are given for the same visible surface point W in the *WCS*, at time slots t and $t + 1$. The task consists in calculating the coordinates of W, where the Z-coordinate of W in the *CCS* is identified with *depth*.

---

[1] Some stabilization of the calculated radii will appear after about eight iterations.

At first, assume that the rotation speed may be under control, i.e. the rotation angle between time t and $t + 1$ is known (KRA). Based on the calibration results, the defined task can be solved. It holds

$$\mathbf{R}W + \mathbf{T} = C_1 \qquad (4.4)$$

where $\mathbf{R}$ denotes the calibrated 3×3 rotation matrix, and $\mathbf{T}$ denotes the calibrated translation vector. For the rotation $\mathbf{R}_A$ of the disc between t and $t + 1$, it holds

$$\mathbf{R}\mathbf{R}_A W + \mathbf{T} = C_2. \qquad (4.5)$$

For the calibrated focal length B and the ideal image points (xP_i, yP_i) at time $t_1 = t$ and $t_2 = t + 1$, it holds that

$$x_{P_i} = \frac{X_{C_i} B}{Z_{C_i}}, \quad \text{and} \quad y_{P_i} = \frac{Y_{C_i} B}{Z_{C_i}}. \qquad (4.6)$$

For the calibrated distortion coefficients κ_1 and κ_1, based on measurements of the distorted image coordinates, at first the ideal image points P_i can be computed, and secondly these ideal points may be used for determining points C_i in the CCS,

$$C_i = \begin{pmatrix} X_{C_i} \\ Z_{C_i} \\ Z_{C_i} \end{pmatrix} = \begin{pmatrix} \dfrac{x_{P_i} Z_{C_i}}{B} \\ \dfrac{y_{P_i} Z_{C_i}}{B} \\ Z_{C_i} \end{pmatrix} = Z_{C_i} \begin{pmatrix} \dfrac{x_{P_i}}{B} \\ \dfrac{y_{P_i}}{B} \\ 1 \end{pmatrix} = Z_{C_i} E_i, \qquad (4.7)$$

by solving the two equation systems in the following Theorem. For abbreviation, let $Z_1 = Z_{C_1}$, and $Z_2 = Z_{C_2}$.

Theorem 4.1 [13]. *For E_i as defined in* (4.7) *it holds*

$$\mathbf{R}^T(Z_1 E_1 - \mathbf{T}) = (\mathbf{R}\mathbf{R}_A)^T(Z_2 E_2 - \mathbf{T}).$$

There are three equations and two unknowns. In fact, the disc rotation angle may be taken as third unknown. But, the equation system "looses its linearity" if considered also for unknown φ_A.

For abbreviation, let $\mathbf{a} = (a_X, a_Y, a_Z) = \mathbf{R}^T E_1$, $\mathbf{b} = (b_X, b_Y, b_Z) = \mathbf{R}^T \mathbf{T}$, and $\mathbf{c} = (c_X, c_Y, c_Z) = \mathbf{R}^T E_2$. Then it holds

Theorem 4.2 [13]. *For*

$$\varphi_A = 2 \arctan\left(\frac{c_2(a_Y b_Z - b_Y a_Z) + c_1(a_X b_Z - b_X a_Z)}{c_2(a_X b_Z + b_X a_Z) - c_1(a_Y b_Z + b_Y a_Z)}\right),$$

$$Z_1 = \frac{b_Z(c_X - c_X \cos(\varphi_A) + c_Y \sin(\varphi_A))}{a_X b_Z - a_Z b_X \cos(\varphi_A) + a_Z b_Y \sin(\varphi_A)},$$

$$Z_2 = Z_1 \frac{a_Z - c_Z}{b_Z - c_Z}.$$

In the abbreviating rule language, these Theorems may be written as

$$[RO, FRA, KRA; cp]: (calibration\ results,\ local\ displacement) \rightarrow depth \quad (4.8)$$

or even

$$[RO, FRA; cp]: (calibration\ results,\ local\ displacement) \rightarrow (depth,\ rotation\ angle).$$
$$(4.9)$$

The calibration results are very accurate, but computing highly accurate flow vectors or point correspondence is still the open problem for real objects on the rotating disc. For polyhedral objects, features computed by a corner response function were suggested for correspondence analysis [16]. For practical evaluation of these both combining Theorems 4.1 and 4.2, complex synthetic objects were considered. The experiment specifications were as follows:

Input images and ground truth: Synthetic objects (visualized by surface shading) were assumed on a rotating disc in front of a camera. During rotation, several projections were computed, and exact correspondence was assumed. The visualized surface structure was used as qualitative ground truth.

Error measure: For interactive evaluation, the resulting surface was graphically represented (3-D triangulation and shaded surface).

In Fig. 4.1, depth reconstruction is illustrated for a cube, using a derivational algorithm A based on Theorem 4.1 (left), and an other derivational algorithm B based on Theorem 4.2 (right). In the experiments, algorithm A was robust for any 2-D motion of point C_1 into C_2 within the image plane. Algorithm B did not work if the direction of the motion vector (u, v) of point C_1 into C_2 is "nearly parallel" to image rows or image columns, i.e.

$$\frac{u}{v} \gg 1 \quad \text{or} \quad \frac{u}{v} \ll 1. \quad (4.10)$$

So far, no mathematical explanation is available for this "bad behavior" in case of applying Theorem 4.2.

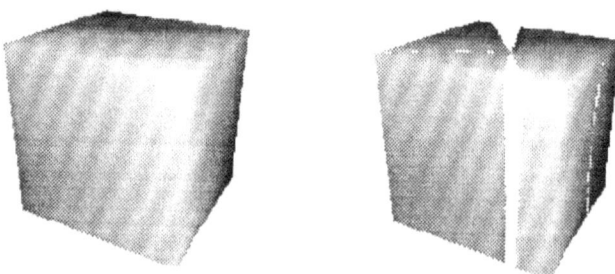

Figure 4.1. Reconstructed depth map of a cube based on Theorem 4.1 (left), and reconstructed depth map based on Theorem 4.2 (right)

As conclusion, a two-step procedure is proposed. At first algorithm B is used for calculating the unique (!) rotation angle:

For some corresponding pairs C_1 and C_2, the rotation angle is calculated.

Then, for the resulting angles a certain mean value is derived as unique rotation angle φ_A.

Then, algorithm A is used to calculate depth values for all pairs of corresponding points C_1 and C_2.

In Fig. 4.2, the reconstruction results are illustrated for a "more complex" synthetic object then the cube of Fig. 4.1. Only two projected images were assumed. The correct motion field was available (motion vectors rounded to pixel positions!), the rotation angle was given (for correct flow vectors, Theorem 4.2 is very robust for calculating this correct rotation angle), and algorithm A was applied.

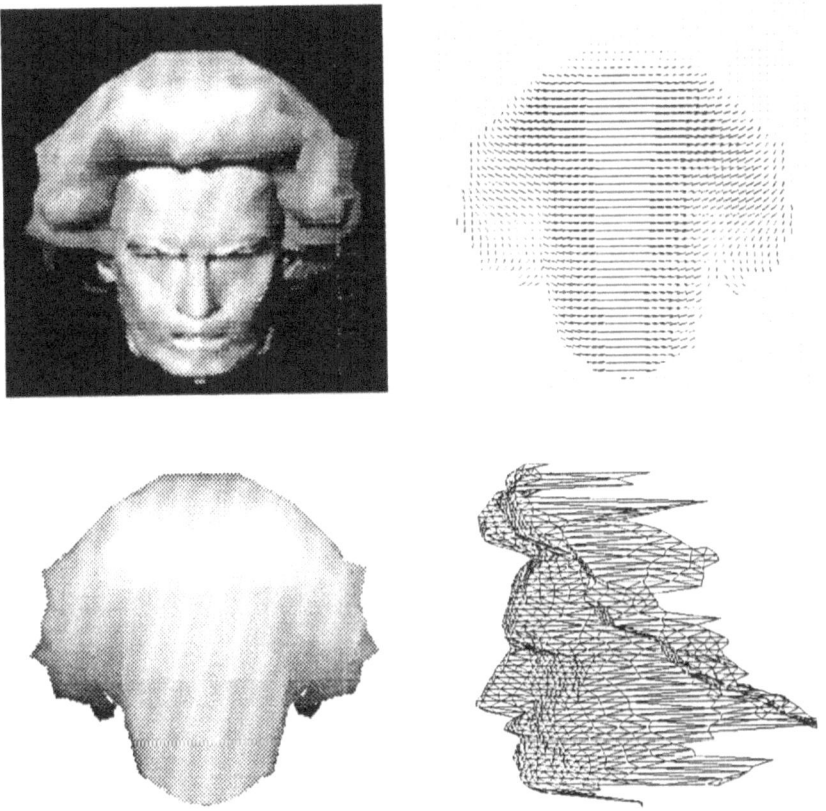

Figure 4.2. Synthetic 3-D object, its motion vector field simulating the rotating disc, the reconstructed depth map based on two (!) projections only, and a 3-D visualization of these depth values as used for interactive error analysis

Because the sketched calibration method of Section 4.1 is very accurate, the same could be qualitatively evaluated for the defined surface reconstruction experiment.

Unfortunately, so far automatic dense flow vector field computation is not available at a quality level allowing similar reconstructions for real objects just by using approach (4.8) or (4.9). By adding noise to ideal motion vector fields it became clear that relatively small distortions will have a great impact on the reconstructed surfaces.

In Fig. 4.3 it is illustrated what "quality" of depth or surface calculation was obtainable. The experiment specifications were as follows:

Input images and ground truth: Objects (as plaster statues, or boxes) were placed on a rotating disc in front of a camera. During rotation, several pictures were taken. The visible surface structure as used as qualitative ground truth.
Error measure: For interactive evaluation, the resulting surface was graphically represented (3-D triangulation and shaded surface).

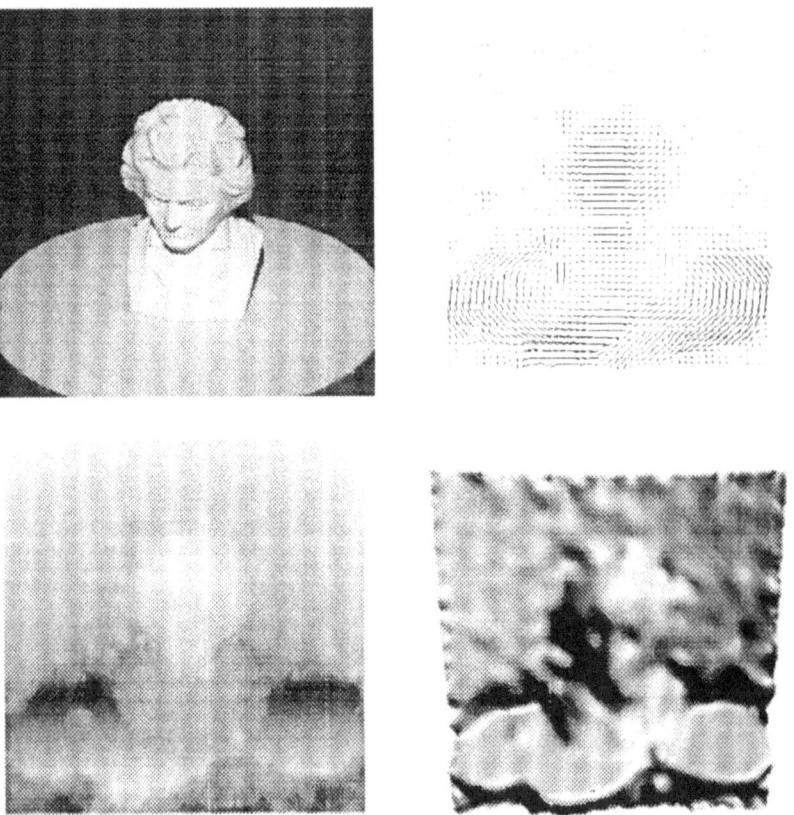

Figure 4.3. A plaster statue on the rotating disc, optical flow field calculated by the method of [3], reconstructed depth map using these flow vectors, and 3-D visualization of this depth map

Comparing many point-based differential methods for computing optical flow, the method by [3] was chosen to behave best for this experiment (By comparing with [4], note that *dense* flow fields were the goal within this experiment.). For 3-D visualization of the depth map, depth values were smoothed. The high error rate of such a motion based surface reconstruction technique is without doubt. For these plaster statues, homogeneous surface textures as assumed in the experiment in Section 2 are not given, leading to "very erroneous" motion fields. Then, the algorithm following rules (4.8, 4.9) will produce "very rough" surface drafts.

Dense optical flow field computation will not lead to complete surface reconstructions within acceptable limits of quality. But, singular correspondences ("sparse flow fields") seem to be realizable for real objects with good accuracy allowing computations of a few 3-D positions of surface points (*fixation points*).

5. Conclusions

Without or with calibration, optical flow or correspondence computation is the critical problem for surface reconstruction if (!) this is based only on motion analysis. Without calibration, also the step from motion to shape possesses very high algorithmic complexity. Many different approaches may be selected, and this paper sketches just a few basic steps, cf. for example [15] for more. A careful evaluation of such approaches, as realized in [8] for sceletonization approaches, should be performed in near future.

If certain a-priori knowledge about object motion is available, a direct approach may be possible without going via shape. At least, this will avoid further sources of errors on the way from intensity images to reconstructed surfaces. Also, dense motion fields are not necessary anymore in such a case if depth is derivable directly from single flow vectors. Thus, a few 3-D fixation points may be calculated by such an approach on object surfaces, and different approaches, as photometric methods may be used "to fill the gaps".

In this paper, several combining theorems were interpreted by qualitative rules of reasoning. This may help to evaluate approaches in the case that they may be considered for integration into a complex computer vision process. Also, by comparing the model assumptions, "gaps" in research, and possible combinations of approaches will be visible. Many detailed questions have to be answered for derivational units, as

— to which extend (conditions, noise etc.) the algorithm is working,
— is it possible to simplify the combining theorem (algebraic expression etc.),
— the numeric stability of the derivational algorithm, its robustness etc.,
— comparisons between derivational algorithms for the same theorem, or
— difficulties in "digitizing" a derivational algorithm.

Also, this paper is incomplete with respect to all known methods relevant to surface from motion. Important modern approaches as David Fleet's phase technique and Goesta Granlund's tensor methods were not yet included into our comparative experimental investigations.

This paper was written to summarize some work realized in the vision group at TU Berlin within the last two years. Especially, results of master thesis projects by Peter Handschack, Dirk Mehren and Volker Rodehorst were included.

References

[1] Aloimonos, J., Shulman, D.: Integration of visual modules: an extension of the marr paradigm. Boston: Academic Press 1990.
[2] Aloimonos, J. (ed.): Active perception. Hillsdale: Lawrence Erlbaum 1993.
[3] Anandan, P.: Measuring visual motion from image sequences. PhD, COINS TR 87-21, Univ. of Massachusetts, Amherst, 1987.
[4] Barron, J. L., Fleet, D. J., Beauchemin, S. S.: Performance of optical flow techniques. Int. J. Computer Vision 12, 43–77 (1994).
[5] Handschack, P., Klette, R.: Quantitative comparisons of differential methods for measuring of image velocity. In: Aspects of visual form processing (Arcelli, C., Cordella, L. P., Sanniti di Baja, G., eds.), pp. 241–250. Singapore: World Scientific 1994.
[6] Handschack, P., Klette, R.: Evaluation of differential methods for image velocity measurement. Comput. Art. Intell. (to appear).
[7] Horn, B. K. P.: Robot vision. New York: McGraw-Hill 1986.
[8] Jaisimha, M. Y., Haralick, R. M., Dori, D.: Quantitative performance evaluation of thinning algorithms in the presence of noise. In: Aspects of visual form processing (Arcelli, C., Cordella, L. P., Sanniti di Baja, G., eds.), pp. 261–286. Singapore: World Scientific 1994.
[9] Kanatani, K.: Group-theoretical methods in image understanding. Berlin, Heidelberg, New York, Tokyo: Springer 1990.
[10] Klette, R.: A framework to computer vision research. Machine Graphics Vision 1, 331–341 (1992).
[11] Klette, R.: Qualitative vision rules, combining theorems, and derivational algorithms. Kibernetika, i Sistemngi Analiz 4, 38–54 (1995).
[12] Klette, R., Rodehorst, V.: Algorithms for shape from shading, lighting direction and motion. Proceed. 5th Int. Conf. CAIP'93, Budapest, September 1993 (Chetverikov, D., Kropatsch, W. G., eds.), pp. 420–427. Lecture Notes in Computer Science vol. 719. Berlin, Heidelberg, New York, Tokyo: Springer 1990.
[13] Klette, R., Mehren, D., Rodehorst, V.: An application of shape reconstruction from rotational motion. Real-Time Imaging 1, 127–138 (1995).
[14] Marr, D.: Vision: a computational investigation into the human representation and processing of visual information. San Francisco: W. H. Freeman 1982.
[15] Maybank, S.: Theory of reconstruction from image motion. Berlin, Heidelberg, New York, Tokyo: Springer. 1993.
[16] Mehren, D., Rodehorst, V.: Gestaltsanalyse komplexer Objekte bei kontrollierter Bewegung. Diplom-Arbeit, TU Berlin, Fachbereich Informatik, 1994.
[17] Muller, D. E.: A method for solving algebraic equations using an automatic computer. Math. Tables Aids Comp. 10, 208–215 (1956).
[18] Nagel, H. H., Enkelmann, W.: An investigation of smoothness constraints for the estimation of displacement vector fields from image sequences. IEEE Trans. PAMI, PAMI-8, 565–593 (1986).
[19] Tsai, R. Y.: An efficient and accurate camera calibration technique for 3D machine vision. Proc. IEEE Conf. Computer Vision and Pattern Rec. 364–374 (1986).
[20] Tsai, R. Y.: A versatile camera calibration technique for high accuracy 3D machine vision methology using off-the-shelf tv camera and lenses. IEEE J. Rob. Automat. 323–344 (1987).

Prof. Dr. R. Klette
Computer Vision Group
Computer Science Department
Technical University at Berlin
FR 3-11, Franklinstrasse 28-29
D-10587 Berlin
Federal Republic of Germany

Computing Suppl. 11, 99–111 (1996)

Properties of Pyramidal Representations*

W. G. Kropatsch, Vienna

Abstract

Properties of Pyramidal Representations. The categorization of different components generalizes the classical concept of image pyramids and provides a powerful tool for efficient image analysis. Three aspects of image pyramids are distinguished: their structure, the contents of their cells and the processes that operate on them. The properties of these three aspects of a pyramidal system are discussed and illustrated by examples. The general theory covers the most recent results, e.g. structure preserving irregular pyramids, sigmoid pyramids, the concept of equivalent interpretation, and the fuzzy curve pyramid.

Key words: Computer vision, image pyramids, hierarchical representations.

1. Introduction

Classical image pyramids have been introduced 1981/82 [8, 25] as a stack of images of decreasing resolutions (Fig. 1). Since then several modifications and additions have been made to the original concept [13] while main properties have been preserved. To structure the many different branches of development and of research we will consider three different aspects of pyramidal representations separately: their structure, the contents of their cells, and the processes that operate on them.

1.1. Topological Structure

The different images in the pyramidal stack are ordered according to their resolution. In this ordering the individual images are called *levels* and are numbered from bottom (highest resolution) to top. Based on the neighborhood relations of an image, the notion of *reduction window* (Fig. 2) is introduced for regular (e.g. square grid) structures. It relates every pixel at a lower resolution to a set of pixels in the level directly below. Often they are called *children*. In a regular pyramid all (interior) cells have the same number of neighbors and children.

* This work was supported by the Austrian National Fonds zur Förderung der wissenschaftlichen Forschung under grant S 7002.

Figure 1. A 3 × 3/2 Gaussian pyramid

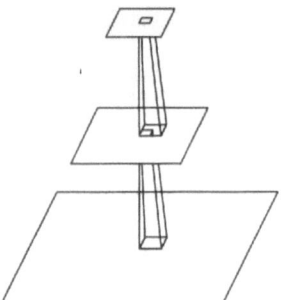

Figure 2. Reduction window

A second parameter that describes different types of pyramids is the *reduction factor*. It captures the rate by which the number of pixels decreases and is described as the quotient of the total number of pixels of two adjacent levels for sufficiently large images (i.e. to avoid boundary effects). In regular pyramids the reduction factor is constant. In the formal notation "2 × 2/4", 2 × 2 defines the size of the reduction window, and the reduction factor 4 expresses that level $k + 1$ contains only a quarter of pixels of level k. The notation (*reduction window*)/ *reduction factor* characterizes the type of the regular pyramid. Figure 3 overlays two adjacent levels of four different types of regular pyramids. Note that adjacent reduction windows may overlap (see Fig. 3b, c, d).

1.1.1. Irregular Pyramids

Newer developments relaxed the regularity of pyramidal structures and used general graph structures instead (Fig. 4). Cells are represented by the vertices (o, ●) of the graph, the vertices corresponding to adjacent cells are joint by edges. In *irregular pyramids* the structure is not defined a priori. There are algorithms like Meer's decimation [20] that make use of some freedom in generating the structure of the pyramid. The lower resolution graph (*level $k + 1$*) is formed by a subset of the *surviving* vertices (●) of the graph. The new edges (*reduced adjacency relations*) are

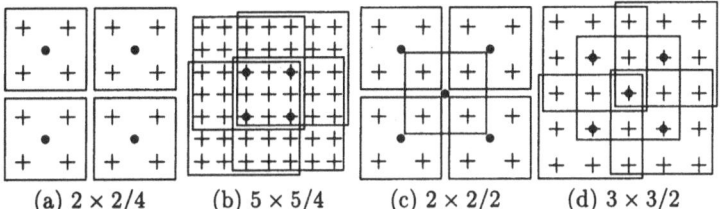

(a) 2 × 2/4   (b) 5 × 5/4   (c) 2 × 2/2   (d) 3 × 3/2

Figure 3. Regular pyramids: levels k (+) and $k + 1$ (●), reduction window

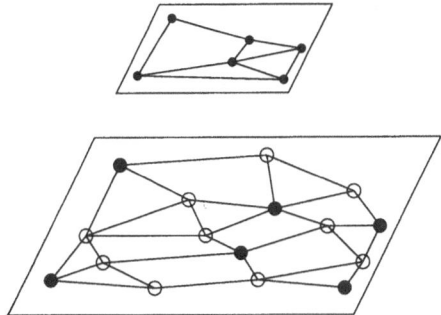

Figure 4. Irregular pyramid

derived from the edges in the level below. Often the surviving vertices should be homogeneously distributed. Meer solved this problem by requiring that the cells of a level in the irregular pyramid form a *maximum independent set* (MIS) of the level directly below. Surviving cells cannot be neighbors in the level below and non-surviving cells must have one surviving neighbor.

More flexibility in choosing the decimation rate can be achieved if surviving cells are allowed to be neighbors. This simplified decimation was proposed in [14]. Its useful properties will be discussed Section 2.1.1.

1.1.2. Dual Pyramids of Planar Graphs

Since images are often projections of the reality into a plane the image structure is planar and can be described by a planar graph. This allows to consider also dual graphs which, in some cases, have useful properties. The dual of a planar graph G can be constructed by (1) placing vertives of the dual graph $\overline{G}$ inside every con-nected region (which is surrounded by a simple closed path) of G and by (2) connecting those dual vertices of which the corresponding regions are adjacent (their boundaries share a common edge of G). Dual pyramids are pairs of pyramids (Fig. 5) where duality exists between corresponding levels. Both regular and irregu-lar pyramids allow construction of the dual pyramid if the underlying neighbor-

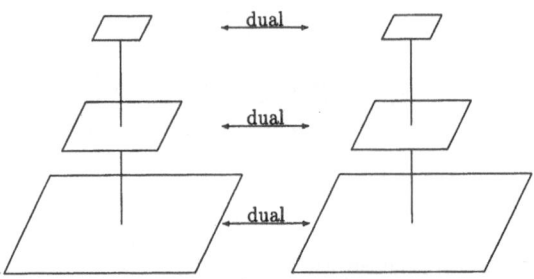

Figure 5. Dual pyramids

hood graphs are planar. Note that this is true for 4-neighborhood but not for 8-neighborhood. In [11] cooperation between pixels and contours efficiently uses $2 \times 2/2$ and $3 \times 3/2$ dual pyramids.

1.2. Cell Contents

The information stored in the cells of a pyramid range from single gray values (Fig. 1) up to very complex descriptions of 'what the cell sees'.

We consider also parameters of one or more models, selected symbols from a vocabulary, simple symbolic descriptions defined by a formal language, and attributed graphs.

Curve pyramids (Fig. 9, top) are an example of a symbolic concept [10]. The part of the curve that is seen through the cell's observation window is described by relating the two parts of the cell's boundary that the curve intersects. Since a square boundary contains only four sides, a finite number of symbols can encode all different ways in which the curve can intersect the square. Connectivity translates into local consistency of symbols and can be expressed as boolean expression of the existence of certain symbols in adjacent cells. Cells in curve pyramids are merged by substitution and by transitive closure.

1.3 Processes

The processes working on pyramidal representations compute or modify either the contents of the cells or the structure of the pyramid.

Reduction functions compute the contents of cells from the contents of cells in the reduction windows. *Refinement functions* carry information from the lower resolutions to the higher resolutions (Fig. 6). Typical functions used are convolutions with predefined (e.g. Gaussian) kernels, filters, interpolation, morphological operations, model fitting (optimization), ... The following two examples demonstrate the wide range of possibilities.

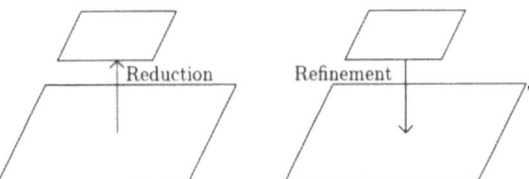

Figure 6. Reduction and refinement functions

Figure 7. Gaussian and Sigmoid pyramids on the road image

Experiments with *sigmoid*-like reduction functions (Fig. 7), which are widely used in neural networks, have been performed. If smoothing has the tendency to reduce the gray level range then the sigmoid stretches the range again and allows better discrimination in some cases (for details see [3]).

For symbolic contents formal grammars can reduce the information contained in the reduction window. An implemented example for this type of process is the *curve pyramid* [10]. Curves are reduced in this concept by a simple process that involves transitive closure. An equivalent chain code ('RULI') representation uses a context sensitive grammar to derive the code of the lower resolution.

The second category of processes modifies the structure of a pyramid. Typical examples are *decimation* processes [20] and *pyramid (re-)linking* [8]. They both create a pyramid, the structure of which is adapted to a static image presented in the base of the pyramid.

Meer's decimation first selects cells that 'survive' to the lower resolution in irregular pyramids. A second process determines the neighborhood relations of the reduced level based on receptive fields.

A new theory, *dual graph contraction* [14], is based on Meer's decimation process and will allow us to derive several useful structural properties. It separates the selection procedure from the pyramid construction. The construction is controlled by a selected subset of vertices, the surviving vertices, and by a selected subset of edges which partition the graph into receptive fields. Dual graph contraction proceeds in two steps: the first step contracts all receptive fields into the surviving vertices. When an edge is contracted, the corresponding dual edge is removed from the dual graph. The second step operates on the dual graph and removes edges that have become self-loops or parallel edges in the first step. Some of these edges are redundant, some contribute to the surviving structure. They can be distinguished in the dual graph: Redundant edges surround faces of degree one and two. Faces of higher degree surround subgraphs that are considered 'important' by initial selection.

To avoid recreation of the complete structure for e.g. every image in a time series, *dynamic modifications* of the initially established structure are studied. Adaptive parent-child relinking allows certain cells to choose "better" parents (e.g. more similar in gray tone) among their possible parents. Recall that adjacent reduction windows may overlap, consequently, cells in the overlap have more than one parent cell. When the parent's cell contents is recomputated from the relinked children a greater homogeneity within the receptive fields and hence a better segmentation is achieved.

Focussing and top-down-segmentation select special parent-child-links that represent a substructure inside the given (irregular or regular) pyramid.

2. Properties

To select appropriate components for a given application their properties must be considered to find an efficient combination that suits the problem at hand. In the following we discuss the properties of the three types of components introduced above separately. Being a property of a single component implies automatically that the resulting overall concept benefits from the advantages of all of its components and that, in most cases, the individual property has its origin in the combined use of structure, representation, and algorithms.

2.1. Structural Properties

If the information in a pyramid is transmitted only across parent-child links every cell c_k at a higher level k summarizes information from cells in the levels below. In the level $k - 1$ directly below the cell c_k they form the reduction window. Every cell of the reduction window in turn links to cells in lower levels. An *equivalent window* covers all cells at a given level $j < k$ that link to the same cell c_k. It defines the domain of measurements that are summarized as contents of c_k.

If we require that every piece of information in the base of the pyramid should have the potential of influencing the summarization process the reduction factor must not be less than the size of the reduction window. If they are the same then every cell has only one parent, the children of cells at the same level form trees that are disjoint. If the reduction factor is less than the size of the reduction window, then equivalent windows of adjacent cells overlap each other, we speake of *overlapping pyramids*. As we will see later this property contributes to the robustness of pyramids [21] although regular pyramids remain position/shift variant [5].

In regular structures the *top-down refinement* of a cell c_k into equivalent windows can be continued recursively. If the size of the cells is reduced (geometrically) according to the reduction factor, the equivalent windows form shapes that converge towards geometrical shapes (squares for $n \times n/4$ type pyramids, octagones for $n \times n/2$ type pyramids) that depend only on the size of cell c_k and the type of pyramid [12]. We call this geometrical region the *receptive field* of cell c_k. Since equivalent windows are monotonically increasing they are all contained in the receptive field. It also characterizes the type of a regular pyramid.

Nacken [23] observed that classical pyramid relinking destroys the topology of the receptive fields in certain cases. He identified the only two configurations when this can happen and developed modifications to the relinking procedure that preserve the topology of receptive fields.

2.1.1. Properties of Irregular Pyramids

The cells of a level in Meer's irregular pyramid form a *maximum independent set* (MIS) of the level directly below. Surviving cells are not neighbors in the level below and non-surviving cells have one surviving neighbor.

Duality between points and faces, and between point-to-point-connections and sides requires planarity of the respective graphs. It could be shown in general that decimation as defined by Meer preserves planarity. Hence the duals of irregular pyramids exist [17] if the base of the pyramid is planar. Furthermore it can be observed and proved that, under certain conditions, the degrees of faces do not grow in higher levels [15]. The latter opens a perspective to implement search mechanisms in the dual concept with bounded neighborhood complexity.

Special cases like vertex degree two and self-loops caused some problems to the use of irregular pyramids until recently. In [27, 14] decimation has been extended to dual graph contraction in a consistent way. It combines operations on both the original and the dual graph. The new algorithm preserves the structure of receptive fields, and it solves the before mentioned special cases of non-decreasing face degrees too.

Figure 8. Receptive fields generated by adaptive dual graph contraction, original and levels of irregular pyramid (from left to right, and top to bottom)

The relaxation of the structural regularity opens the possibility to adapt the structure to the data at hand with the goal: *The same object should create the same structure independent of the location and orientation in the image.* Promising experimental results can be found in [9, 22]. Figure 8 shows the development of receptive fields (displayed by the same gray value) as the pyramid is constructed by dual graph contraction. In five successive contractions the original 16×16 labeled image is reduced to the region adjacency graph. The selection prohibits differently labeled regions to be merged and is stochastic otherwise.

2.2. Representation Properties

Burt [7] has introduced the term "*equivalent weighting function*" to describe the computation of any level of a convolution-pyramid directly from the data in the base. Such equivalent weighting functions allow to derive several properties that higher levels in the pyramid have, e.g. the low-pass character of Gaussian type kernels. The notion "*equivalent interpretation*" extends Burt's term to complex descriptions that may also be in symbolic form. It relates properties of the base with properties in higher levels. In the example of the $2 \times 2/2$ curve pyramid, following properties are preserved [10]:

- The connectivity of a sequence of curve segments is preserved up to the level of annihilation of the curve, e.g. when the curve is completely covered by one pyramidal cell.

Figure 9. A 2 × 2/2 curve pyramid

- If there is no curve connecting two sides of a cell at any level of the pyramid, then there is no curve connecting the corresponding sides in any lower level cell.
- The length of a curve description does not increase in higher levels. This property has been used to filter out short (noisy) curve segments (Fig. 9).

The *fuzzy curve pyramid* [4] is an extension integrating the magnitude of individual curve segments into the general framework. The *strength* of a curve is defined as the smallest magnitude of the curve's segments. The reduction function of the fuzzy curve pyramid maintains the strongest curves.

Dual pyramids allow cooperation between complementary types of representation. In Figures 10 and 11 the gray level reduction process does not smooth across long boundaries of high contrast remaining in the dual curve pyramid after removing the short curve segments [11].

The complexity of data structures and search processes requires a bounded length of the description in a cell. The enforcement of this constraint automatically introduces *abstraction* and generalization if all the information of the reduction window cannot be described in one single cell.

Recognition and identification would be simple if the same object would always generate the same description independent of the appearance in the image. That introduces invariance requirements not common to image analyses processes: transformations between the imaging geometry and the object like shift, rotation, scaling, and perspective often influence the resulting interpretation. A pyramidal

Figure 10. A 3 × 3/2 Gaussian pyramid using the curve information from the 2 × 2/2 curve pyramid

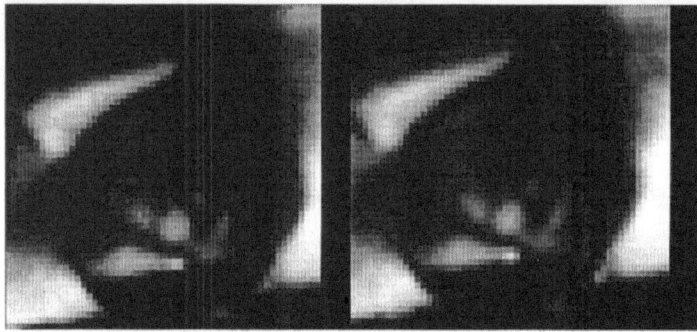

Figure 11. Zoomed part of Gaussian pyramid (left) and Gaussian pyramid with curve information

cell c can compute many properties (e.g. mean gray value) of parts of the image that fall within the receptive field of c. Information outside of the receptive field of c is not "visible" for c and, hence, may cause varying results depending on the location of the part within the image window. Solutions to the invariance problem are:

- *Active vision* forces objects into a specific location of the imaging window (see [2]).
- An *adaptive structure* may allow the system to build the same structure wherever the object is located (e.g. in adaptive irregular pyramids).
- The recovery of a general model may *adapt free parameters* to the given imaging situation [18, 19].

2.3. Processing Properties

In any regular pyramid the number of neighbors is fixed. As a consequence any search in the neighborhood of a cell has constant complexity independent of the type of pyramid. All processes using only the contents of adjacent cells are *local*. In most cases the computation of a pyramidal cell uses only data from its parents or from its children. Hence all processes at one level may work simultaneously.

Both reduction and refinement propagate data up and down in the pyramid. Hence their computational complexity depends only on the number of levels in the pyramid, e.g. on the log(*image − diameter*).

This computational efficiency may even affect sequential processing: It can be shown that the recursive (sequential) computation of the n'th level needs less operations than direct computation by equivalent weighting functions.

Other processes benefit from a reduction of search space as a consequence of shorter paths at higher pyramid levels. Many examples can be enumerated that take advantage of this property: blob detection [6], stereo matching, texture analysis, and tracking [1], generation of attractors of iterated function systems (IFS) [16], speeding up Cornwall's life game (Up to 50% speed up could be achieved in practical experiments).

The efficiency of pyramids is drastically reduced if the neighborhood complexity is no more constant, e.g. if a quarter of all cells of a level have to be processed sequentially. In irregular pyramids the degree of a vertex (which corresponds to the size of the neighborhood) cannot be bounded. Search processes in dual irregular pyramids benefit from the fact that the degrees in the dual graph remain bounded by the (face-)degrees in the base. As a recent result it could be proven in [26] that the building of a dual irregular pyramid takes on the order of $\mathcal{O}\{(\log D_{max})^2 \cdot \log(image - diameter)\}$ parallel steps where D_{max} is the maximum degree of a vertex.

2.4. Limitations

Most of the above advantages assume implicitly that the data have a great amount of redundancy and that the output of the system is small in relation to the vast amount of input data. However if the result involves a global exchange of information among the input data structure (e.g. a permutation of the input pixels) then the pyramid does not provide a *large enough bandwidth* for this type of exchange. In such case missing connections between cells must be compensated by increased computational complexity.

3. Conclusion

Rosenfeld [24] exposed the properties of two "variable-resolution representations" in 1982: pyramids and quadtrees. Since then different reduction functions have greatly enhanced the computational power of pyramids. However in the pure bottom-up processing mode, *pyramids do not compute anything new*, which could not be computed from the cells of the receptive field. The notion of *equivalent interpretations* also expresses this fact. Why then use pyramids any more?

- Pyramids decompose an often very complex computation into a few relatively simple processing steps.
- Global operators decompose into a few local operators which can be applied in parallel.
- Pyramids are computationally extremely efficient.
- The limitation of a cell's storage capacity enforces abstraction, less important data are neglected in higher levels.
- Exterior knowledge sources can be integrated at different levels of abstraction. It could define the language of interpretation.
- Irregular pyramids adapt their structure to the data rather than the parameters of the underlying models.
- The graph structures of irregular pyramids provide a smooth transition to mid- and high-level vision.
- Dual pyramids combine region and boundary type representation in a unified concept.
- The interpretation of dual decimation in terms of cellular complexes open the field to the consistent treatment of higher dimensions.

Acknowledgements

The author is grateful to Horst Bischof for several interesting discussions and for running several experiments reported in this paper. Dieter Willersinn contributed several ideas to the dual graph contraction and Herwig Macho implemented the program and produced the first results.

References

[1] Ackermann, F., Hahn, M.: Image pyramids for digital photogrammetry. In: Digital photogrammetric systems (Ebner, H., et al., eds.), pp. 43–58. Karlsruhe: Wichmann 1991.
[2] Aloimonos, Y., ed.: Active Perception. Hillsdale, New Jersey: Lawrence Erlbaum 1993.
[3] Bischof, H.: Pyramidal Neural Networks. PhD thesis, Technische Universität Wien, 1993.
[4] Bischof, H., Kropatsch, W. G.: The fuzzy curve pyramid. In: 12th IAPR International Conference on Pattern Recognition Vol. 1 (Peleg, S., Ullman, S., Yeshurun, Y., eds.), pp. 505–509. Washington, Brussels, Tokyo: IEEE Comp. Soc., 1994.
[5] Bister, M., Cornelis, J., Rosenfeld, A.: A critical view of pyramid segmentation algorithms. Pattern Rec. Lett. *11*, 605–617 (1990).
[6] Blanford, R. P., Tanimoto, S. L.: Bright spot detection in pyramids. Comput. Vision Graphics Image Proc. *43*, 133–149 (1988).
[7] Burt, P. J., Adelson, E. H.: The Laplacian pyramid as a compact image code. IEEE Trans. Comm. *COM-31*, 532–540 (1983).
[8] Burt, P. J., Hong, T.-H., Rosenfeld, A.: Segmentation and estimation of image region properties through cooperative hierarchical computation. IEEE Trans. Systems Man Cybernetics *SMC-11*, 802–809 (1981).
[9] Jolion, J.-M., Montanvert, A.: The adaptive pyramid, a framework for 2D image analysis. Comput. Vision Graphics Image Proc. Image Underst. *55*, 339–348 (1992).
[10] Kropatsch, W. G.: Curve representations in multiple resolutions. Pattern Rec. Lett. *6*, 179–184 (1987).
[11] Kropatsch, W. G.: Preserving contours in dual pyramids. Proc. 9th International Conference on Pattern Recognition, pp. 563–565, Rome, Italy, November 1988. IEEE Comp. Soc.
[12] Kropatsch, W. G.: Rezeptive Felder in Bildpyramiden. In: Mustererkennung 1988 (Bunke, H., Kübler, O., Stucki, P., eds.), pp. 333–339. Berlin, Heidelberg, New York, Tokyo: Springer 1988 (Informatik Fachberichte Vol. 180).

[13] Kropatsch, W. G.: Image pyramids and curves—an overview. Technical Report PRIP-TR-2, Institute f. Automation 183/2, Dept. for Pattern Recognition and Image Processing, TU Wien, Austria, 1991.

[14] Kropatsch, W. G.: Building irregular pyramids by dual graph contraction. Technical Report PRIP-TR-35, Institute f. Automation 183/2, Dept. for Pattern Recognition and Image Processing, TU Wien, Austria, 1994.

[15] Kropatsch, W. G., Montanvert, A.: Irregular versus regular pyramid structures. In: Geometrical problems of image processing (Eckhardt, U., Hübler, A., Nagel, W., Werner, G., eds.), pp. 11–22. Berlin: Akademie Verlag 1991.

[16] Kropatsch, W. G., Neuhauser, M. A., Leitgeb, I. J.: Iterated function systems—A direct discrete approach with pyramids. In: Pattern recognition 1992, pp. 108–118. OCG-Schriftenreihe, Band 62. Österr. Arbeitsgemeinschaft für Mustererkennung. München: R. Oldenburg 1992.

[17] Kropatsch, W. G., Reither, C., Willersinn, D., Wlaschitz, G.: The dual irregular pyramid. In: Proceedings CAIP'93, Budapest, pp. 31–40, 1993.

[18] Leonardis, A.: Image analysis using parametric models: model-recovery and model-selection paradigm. PhD thesis, University of Ljubljana, Department of Computer and Information Science, Faculty of Electrical Engineering and Computer Science, 1993.

[19] Leonardis, A.: A robust approach to estimation of parametric models. In: Theoretical foundations of computer vision (Kropatsch, W., Klette, R., Solina, F., eds.), pp. 113–130. Wien, New York: Springer 1996 (Computing Suppl. 11).

[20] Meer, P.: Stochastic image pyramids. Comput. Vision Graphics Image Proc. 45, 269–294 (1989).

[21] Meer, P., Jiang, S.-N., Baugher, E. S., Rosenfeld, A.: Robustness of image pyramids under structural perturbations. Comput. Vision Graphics Image Proc. 44, 307–331 (1988).

[22] Montanvert, A., Meer, P., Rosenfeld, A.: Irregular tesselation based image analysis. In: Proc. 10th International Conference on Pattern Recognition, pages 474–479, Atlantic City, New Jersey, USA, June 1990. IEEE Comp. Soc. Vol. I.

[23] Nacken, P. F.: Image analysis methods based on hierarchies of graphs and multi-scale mathematical morphology. PhD thesis, Universiteit van Amsterdam, The Netherlands, June 1994.

[24] Rosenfeld, A.: Quadtrees and pyramids: hierarchical representation of images. Technical Report TR-1171, University of Maryland, Computer Science Center, May 1982.

[25] Rosenfeld , A., Kak, A. C.: Digital picture processing, 2nd ed., Vol. 1 and 2. New York: Academic Press 1982.

[26] Willersinn, D.: Parallel graph contraction for dual irregular pyramids. Technical Report PRIP-TR-28, Institute f. Automation 183/2, Dept. for Pattern Recognition and Image Processing, TU Wien, Austria, 1994.

[27] Willersinn, D., Kropatsch, W. G.: Dual graph contraction for irregular pyramids. In: 12th IAPR International Conference on Pattern Recognition, Vol. III (Peleg, S., Ullman, S., Yeshurun, Y., eds.), pp. 251–256. Washington, Brussels, Tokyo: IEEE Comp. Soc., 1994.

Dr. W. G. Kropatsch
Institute for Automation 183/2
Department for Pattern Recognition
and Image Processing
Technical University of Vienna
Treitlstr. 3
A-1040 Vienna
Austria
e-mail: krw@prip.tuwien.ac.at

Computing Suppl. 11, 113–130 (1996)

A Robust Approach to Estimation of Parametric Models*

A. **Leonardis**, Ljubljana

Abstract

A Robust Approach to Estimation of Parametric Models. This article presents a robust method for estimation of parametric models. The method consists of two procedures: model-recovery and model-selection. The model-recovery procedure systematically recovers a redundant set of parametric models in a local-to-global fashion, iteratively combining data classification and parameter estimation. The model-selection procedure, defined as a quadratic Boolean problem, then searches for the subset of the recovered models which produce the simplest global description. To achieve a computationally efficient method the model-recovery and the model-selection are combined in an iterative way. The main features of the method are a high degree of resistance to outliers and the insensitivity to incorrect initial estimates. The method has been successfully applied to linear as well as nonlinear parameter estimation problems, e.g. for recovering variable-order bivariate polynomials and superquadric models in range images, and parametric curve models in edge images.

Key words: Robust estimation, combinatorial optimization, segmentation, minimum description length, parametric models, surface fitting, curve fitting, superquadrics.

1. Introduction

The characteristics of the visual signal, such as the overwhelming amount of data, high redundancy, relevant information clustered in space and time, indicate that certain organization and aggregation principles have to be utilized to reduce the computational complexity of the visual processes and to bridge the gap between the raw data and symbolic descriptions. In computer vision, the data aggregation problem is most commonly approached by fitting models of visual phenomena to image data. Since the image data is inherently unreliable [18], a fitting method should be able to cope both with

- noise that is well-behaved in a distributional sense and
- outliers, which are either large measurement errors or data points belonging to other distributions (models).

It is well known that the least-squares estimator which is based on the assumption of a pure Gaussian noise is very sensitive to outliers in the data set which may lead to extremely poor results.

---

* This work was supported in part by The Ministry for Science and Technology of The Republic of Slovenia (Projects P2-1122 and J2-6187).

In Schunck [18] it is argued that visual perception is fundamentally a problem in discrimination: data must be combined with like data and outliers must be rejected. In other words, vision algorithms must be able to combine data while simultaneously discriminating between data that should be kept distinct, such as outliers (errors) and data from other regions. In fact, the data association problem (grouping) makes the task of machine perception fundamentally different from the traditional estimation problems.

The estimators that remain stable in the presence of various types of noise and can tolerate a certain portion of outliers are known under the generic name of *robust estimators* [6, 17]. These estimators are characterized by the concepts of *efficiency* and *breakdown point*. Efficiency refers to the relative ability of an estimator to yield optimal estimates at the assumed noise distribution. The breakdown point of an estimator is determined by the smallest portion of outliers in the data set at which the estimation procedure can produce an arbitrarily wrong estimate. It is especially the aspect of the breakdown point which is usually emphasized in the design of robust estimators for vision algorithms. This indicates that the use of robust estimators in computer vision is often much more in the function of rejecting outliers than in the function of optimally estimating parameters of the models in the case of non-Gaussian data distributions that may arise from the nature of the physical data.

Some robust local operators [2] can theoretically achieve the maximum breakdown point of 0.5 which means that the estimate remains unchanged if less than half of the data are outliers. In practice, the breakdown point may be much lower. Li [13], for example, has shown that various classes of robust local operators have break-down points that are less than $1/(p + 1)$, where p is the number of parameters in the regression. Moreover, Kim et al. [7] have pointed out the infeasibility of using robust local operators in the transition area of two (or more) statistical populations (models). Besides many of these methods are sensitive to initial estimates and computationally very expensive. However, their performance can be improved when they are used as building blocks in a more complex method.

In this paper we present one such method for robust estimation of parametric models. The method can be considered as a general framework in which different parameter estimation techniques can be embedded. The method consists of two procedures: model-recovery and model-selection. The first procedure ensures data consistency by iteratively combining data classification and parameter estimation, enabling the rejection of outliers. The overall insensitivity to initial estimates is achieved by systematically recovering a redundant set of parametric models and then passing them to the model-selection procedure which selects a subset of these models based on their spatial extent, goodness-of-fit, and number of parameters.

The paper is organized as follows. In Section 2 we point out some of the features of the robust estimation methods. In Section 3 we outline the recover-and-select paradigm. In Section 4 we show how the method can be used to recover variable-order bivariate polynomials and superquadric models in range images, and parametric curve models in edge images. In conclusion we summarize our paradigm and outline the work in progress.

2. Robust Estimation Methods

In this section we mention some of the robust estimation methods and point out some of their features. Robust window operators [2, 7, 14, 19] have been brought into computer vision as an answer to the problems encountered when using standard least-squares methods in windows containing outliers or more than one statistical population. M-estimators, for example, tackle the problems by either rejecting "bad" data points from the calculation of model estimate (*hard redescenders*) or down-weighting their influence on the final result (*soft redescenders*). This is achieved by first computing an initial estimate and then refining it by repeated re-weighting of the data points. Several disadvantages can be identified with this approach. Since the information obtained in one window is usually *not* shared among the neighboring windows, the parameter estimation has to start in each window from scratch, i.e., there is no a priori information which data points could, or could not, belong to the model. As the weighting of the data depends on the parameters and vice versa, there is no closed-form solution, and an iterative procedure has to be used for parameter estimation. Another consequence of such local approach is that the number of outliers that can be tolerated successfully is limited. Also the initial estimates are unreliable since they are based on a small number of data points in a window (if we enlarge the size of the window, we increase the risk of encompassing data that belong to several different models). For certain redescending (nonmonotonic) functions there are no guarantees of convergence to a unique solution, which means that the method may not be able to correct the initial errors.

In general, the size of the window plays an important role in the case of robust window operators since it influences the accuracy of the solution and the computational time. Many robust operators are hard to implement and computationally demanding even for neighborhoods of reasonable sizes [14].

To overcome some of these problems, methods were designed which estimate parameters of the models on domains that are not restricted to prespecified windows [1, 3, 4]. While robust window operators have been designed to tolerate a certain portion of outliers, these methods try to prevent the outliers to enter the parameter estimation procedure in the first place[1]. The main idea is that once we have an initial model, we can propagate the information to the neighboring re-

---

[1] In this aspect these methods are related to the classical problem of computer vision, namely the *segmentation* which is most commonly defined as partitioning the data into sets that correspond to single statistical populations.

gions. Data points that are found to be consistent with the model are sequentially added, outliers are rejected, and the parameters of the model are re-computed. Depending on the weighting function of the estimator, the new data points that are iteratively added to the model can also be properly weighted [15].

The main problem that remains unsolved with these approaches is how to determine an initial set of data points (a seed) that would yield a reliable estimation of initial model parameters. The set of initial data points must be kept small to reduce the probability that the data points would belong to multiple models. As a consequence, the estimation of initial model parameters is unreliable. We can argue that neither extensive preprocessing nor the usage of robust estimators can prevent some of the initial estimates from being incorrect or too crude to be used to guide the data aggregation process.

The other problem is the influence that the developing models have on one another. It has been a common practice to develop models in a sequence and those recovered earlier in the process constrained the ones developed later by restricting their domains only to unclassified data points. Erroneous results are thus propagated through the entire procedure resulting in an overall faulty output.

In the paradigm, which is described in the next section, we made an attempt to develop a method that can cope with these problems.

3. Recover-and-Select Paradigm

In this section we present the recover-and-select paradigm. First we describe the *model-recovery* procedure which includes the strategy for grouping image elements and estimating parameters of the models. We also discuss the selection of the seeds. Then we explain how the models with the best descriptive power are selected via optimizing the quadratic Boolean problem. Finally we describe our approach in combining model-recovery and model-selection in an iterative way, which substantially reduces the computational complexity of the method.

Perhaps closest in spirit to our approach is the work done by Pentland [16] who developed an image segmentation system which first creates a large set of hypotheses about the scene part structure (in terms of binary templates) and then uses a modified Hopfield-Tank network that searches for the subset of these hypotheses which constitute the most likely description of the image.

3.1. Model Recovery

We solve the problem of data classification and parameter estimation by an iterative approach, which is conceptually similar to the one described by Besl [1] and Chen [3]. However, the seed selection and the independent recovery of individual models make our approach fundamentally distinct. This leads to a variety of unique features such as redundancy, possible parallelism, and enables provable termination of the algorithm without sacrificing the reliability of the method. Besides, our method treats the recovered models only as *hypotheses* which then compete to be selected in the final description.

3.1.1. Seed Selection

As we have discussed, the selection of the seeds has a major effect on the success or failure of the overall procedure. Chen [3] proposed that a window is moved around in an image searching for an adequate amount of data that is statistically consistent in the sense that the data points belong to the same model. Thus, the requirement of classifying all points from a certain model is relaxed to finding only a small subset. On the other hand, Besl [1] followed the approach of extensive preprocessing to determine the initial estimates. The preprocessing involved computing second-order properties in the local neighborhood of every pixel which is noise sensitive. In the RANSAC paradigm [4] the algorithm randomly selects a minimal set of points necessary to fit a model. The resulting model is then tested for validity, which is based on the number of data points in the consensus set of the model.

We argue that despite using elaborate techniques there is no guarantee that every seed will lead to a good result. For example, the seed consistency can sometimes be satisfied on low strength C^0 and C^1 discontinuities. As a remedy we propose an approach which is not sensitive to the selection of inappropriate seeds. The idea is to *independently* build *all possible* models using all statistically consistent seeds, found in a grid of windows overlaid on the image, and then use the recovered models as hypotheses that could compose the final result.

To determine the statistical consistency of a seed, we fit a model[2] to the data points in the seed window. We take the seed as statistically consistent if the goodness-of-fit indicates that there is a high probability that all the data points in the seed window belong to the same model. For each statistically consistent seed we estimate the parameter(s) of the model and proceed with the model-recovery procedure which simultaneously combines data classification and model fitting.

3.1.2. Data Classification and Model Fitting

The procedure for the recovery of parametric models can be partitioned into three distinct modules. A schematic diagram of the procedure is shown in Fig. 1.

---

[2] In the case that we have a hierarchy of models of increasing complexity we always start with the hypothesis that the initial data points belong to the simplest model, i.e., the one with the minimum number of parameters. This is due to the limited amount of reliable information that can be gathered in an initially local area.

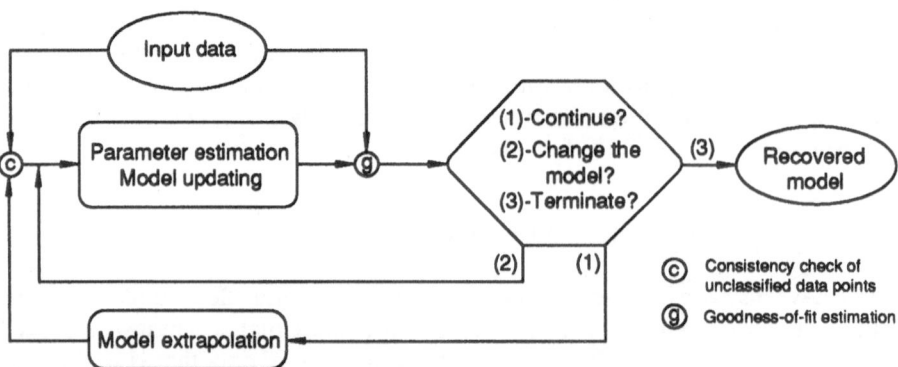

Figure 1. A schematic diagram outlining the model-recovery procedure

Data point classification: Let us suppose that we have a partially recovered model. An efficient search for more compatible points is performed by extrapolating the current model and testing the new data points for consistency. The consistency check can involve sophisticated methods to establish the compatibility of a data point to the model, or a simple distance measure together with a threshold can be used to classify the data points into inliers and outliers. Also, additional constraints can be employed, for example to preserve the topological properties of the models (connectivity). Outliers, which can be classified either as extreme measurement deviations or as data points that belong to neighboring models, are rejected. New consistent image elements are temporarily included in the data set and passed to the parameter estimation module.

Parameter estimation: For the given set of data points, the type of the parametric model, and the estimation technique we compute the parameters of the model and the goodness-of-fit.

Decision making: If sufficient similarity is established between the model and the data (goodness-of-fit), we accept the currently estimated parameters, together with the current data set, and proceed with a search for more compatible points. Otherwise a decision is made whether to replace the currently used model with a more complex one (if there is one), calculate a new set of parameters, and re-evaluate the goodness-of-fit, or terminate the procedure. The procedure terminates if the goodness-of-fit does not improve significantly despite using a higher-order model, or if no higher-order models are available.

The output of the model-recovery procedure for the i-th model consists of three terms:

1. The region R_i, which represents the domain of the model and encompasses $n_i = |R_i|$ image elements that belong to the model,
2. the set of model parameters $\mathbf{a}_i$ (let N_i denote the cardinality of this set), and
3. the error-of-fit measure ξ_i which evaluates the conformity between the data and the model.

While we presented the entire procedure on a general level, i.e., independent of particular types of models and parameter estimation techniques, specific procedures designed to operate on individual types of models can differ significantly. Different parameter estimation techniques are due to different dependencies of fitting functions on the set of unknown parameters. For example, a linear or a nonlinear parameter estimation procedure can be a direct consequence of the choice in defining the measure of the distance between the model and the data.

Since the model-recovery is performed for all statistically consistent seeds, the complete output of the above procedure consists of numerous recovered models, which represent the candidates for the final description of the data.

3.2. Model Selection

The redundant set of parametric models obtained by the model-recovery procedure is a direct consequence of the decision that a search for parametric models is systematically initiated everywhere in an image. Several of the models are partially or completely overlapped. The task of obtaining a subset of the recovered models is defined as a selection procedure which considers many competitive solutions and selects those models that produce the simplest global description. The principle of simplicity has a long history in psychology (Gestalt principles), and the formalization of this principle led in information theory to the method of *Minimum Description Length* MDL, which has recently found its way to computer science, including computer vision [5, 8, 16]. According to the MDL principle, those models are selected from the set of recovered models that describe the data with the shortest possible encoding. As it will be shown in the next subsection, models that provide an efficient encoding should encompass a large number of data points and have a high value of the goodness-of-fit measure. These are exactly the characteristics of the models that contain only the data points belonging to a single statistical population. In contrast, models that incorporated outliers or contain data points that belong to a mixture of data populations either accumulate substantial errors (which prevent them from growing further) or fail to find more compatible points. In any case, these models span a relatively small number of data points and have a poor goodness-of-fit.

In the next two subsections we describe the objective function which encompasses the information about the competing models and the optimization procedure which selects a set of models, respectively.

3.2.1. Objective Function

We want to describe parts of the image, or possibly the whole image, in terms of a selected subset of the set of all recovered models. Let vector $\mathbf{m}^T = [m_1, m_2, \ldots, m_M]$ denote a set of models, where m_i is a *presence-variable* having the value 1 for the presence of the model and 0 for its absence in the final description, and M is the number of all models. The length of encoding of an image L_{image} can be given as the sum of two terms

$$L_{\text{image}}(\mathbf{m}) = L_{\text{pointwise}}(\mathbf{m}) + L_{\text{models}}(\mathbf{m}). \tag{1}$$

$L_{\text{pointwise}}(\mathbf{m})$ is the length of encoding of individual data points that are not described by any model, and $L_{\text{models}}(\mathbf{m})$ is the length of encoding of data described by the selected models. The idea is to select a subset of models that would yield the shortest length description.

We can translate Eq. (1) into our particular case using the outcome of the model-recovery procedure

$$L_{\text{image}}(\mathbf{m}) = K_1(n_{\text{all}} - n(\mathbf{m})) + K_2 \xi(\mathbf{m}) + K_3 N(\mathbf{m}), \tag{2}$$

where n_{all} denotes the number of all data points in the input and $n(\mathbf{m})$ the number of data points that are explained by the selected models. $N(\mathbf{m})$ specifies the number of parameters which are needed to describe the selected models, and $\xi(\mathbf{m})$ gives the deviation between the models and the data that these models describe. K_1, K_2, K_3 are weights which can be determined on a purely information-theoretical basis (in terms of bits), or they can be adjusted in order to express the preference for a particular type of description [9].

Now we can state the task as follows: Find $\hat{\mathbf{m}}$ such that

$$L_{\text{image}}(\hat{\mathbf{m}}) = \min_{\mathbf{m}} L_{\text{image}}(\mathbf{m}). \tag{3}$$

Since n_{all} is constant, minimization of Eq. (2) is equivalent to maximizing the expression

$$F(\hat{\mathbf{m}}) = \max_{\mathbf{m}} F(\mathbf{m}) = K_1 n(\mathbf{m}) - K_2 \xi(\mathbf{m}) - K_3 N(\mathbf{m}). \tag{4}$$

This equation supports our intuitive thinking that an encoding is efficient (description is simple) if the number of data points described by a model is large, the deviations low, and the number of model parameters small.

So far, the optimization function has been discussed on a general level. More specifically, our objective function which takes into account the individual models has the following form:

$$F(\mathbf{m}) = \mathbf{m}^T \mathbf{Q} \mathbf{m} = \mathbf{m}^T \begin{bmatrix} c_{11} & \cdots & c_{1M} \\ \vdots & & \vdots \\ c_{M1} & \cdots & c_{MM} \end{bmatrix} \mathbf{m}. \tag{5}$$

The diagonal terms of the matrix $\mathbf{Q}$ express the cost-benefit value for a particular model M_i

$$c_{ii} = K_1 n_i - K_2 \xi_i - K_3 N_i, \tag{6}$$

while the off-diagonal terms handle the interaction between the overlapping models

$$c_{ij} = \frac{-K_1|R_i \cap R_j| + K_2\xi_{i,j}}{2}, \tag{7}$$

$|R_i \cap R_j|$ is the number of points that are explained by both models, and $\xi_{i,j}$ is defined as

$$\xi_{i,j} = \max\left(\sum_{R_i \cap R_j} d_{M_i}^2, \sum_{R_i \cap R_j} d_{M_j}^2\right). \tag{8}$$

The error terms $d_{M_i}^2$ and $d_{M_j}^2$ are calculated in the region of intersection $R_i \cap R_j$ and correspond to deviations of the data from the i-th and j-th model, respectively.

The objective function takes into account the interaction between different models which may be completely or partially overlapped. However, like Pentland [16], we consider only the pairwise overlaps in the final solution.

From the computational point of view, it is important to notice that the matrix $\mathbf{Q}$ is symmetric, and depending on the overlap of the models, it can be sparse or banded. All these properties of the matrix $\mathbf{Q}$ can be used to reduce the computations needed to calculate the value of $F(\mathbf{m})$.

3.2.2. Solving the Optimization Problem

We have formulated the problem in such a way that its solution corresponds to the global extremum of the objective function. Maximization of the objective function $F(\mathbf{m})$ belongs to the class of combinatorial optimization problems (quadratic Boolean problem). Since the number of possible solutions increases exponentially with the size of the problem, it is usually not tractable to explore them exhaustively. Thus the exact solution has to be sacrificed to obtain a practical one. Various methods have been proposed for finding a "global extreme" of a class of nonlinear objective functions. However, they are in general too time consuming to be applicable in most situations. It turned out that in our case, for well-behaved inputs (in the sense of being well describable by the chosen set of models), we obtain reasonable solutions by a direct application of the *greedy algorithm*. The algorithm is simple: We start with the state in which no models have been selected. The initial value of $F(\mathbf{m})$ is 0. A successor state is formed by adding a not yet selected model to the current description. The selected model is the one which contributes the most to the value of the objective function. The process of adding models to the current description is repeated as long as the value of the objective function can be increased.

In other words, the models are selected in the sequence that corresponds to the size of their contributions to the objective function, which is equivalent to applying at

A. Leonardis

each stage of the algorithm the *winner-takes-all* principle. A discussion on when the greedy algorithm produces satisfactory results and on the time complexity of the model-selection procedure is given in [9].

3.3. Model-Recovery and Model-Selection

After having explained the two major components of our method, namely the module for model recovery and the module for model selection, we now describe how they can be combined to obtain a fast and efficient overall method. For example, the modules can be applied in succession, as shown in Fig. 2a. All the models are first grown to their full extent and then passed to the selection module. We call this approach *Recover-then-Select*. However, the computational cost of growing all the models completely is prohibitive in most cases. Instead it would be desirable to discard some of the redundant and superfluous models even before they are fully grown. This suggests incorporating the selection procedure into the recovery procedure, as shown in Fig. 2b. We call this approach *Recover-and-Select*. The recovery of currently active models is interrupted by the model-selection procedure which selects a set of currently optimal models which are then passed back to the model-recovery procedure. This process is repeated until the remaining models are completely recovered. The trade-offs which are involved in the dynamic

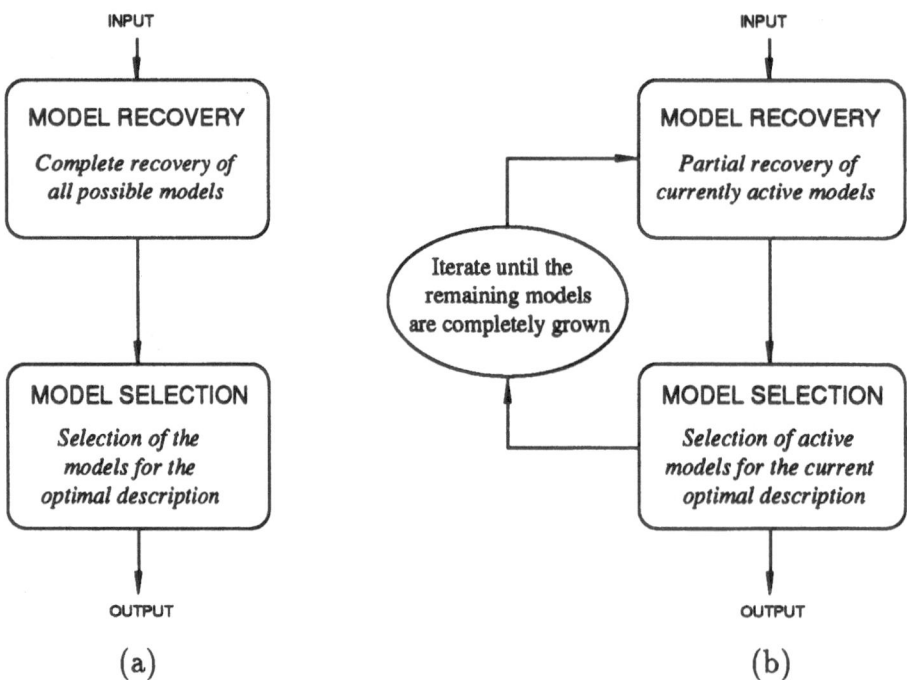

Figure 2. Model recovery and selection: **a** Recover-then-Select, **b** Recover-and-Select

Table 1. *Recover-and-Select* algorithm

---

input: an image

determine a set of seeds

for all seeds **do**
   fit a model (estimate parameters)
   if (GOF = = OK) put the model into the set of currently *active, not fully grown*
models
   else the seed is rejected

while there are any *active, not fully grown* models **do**
   for all *active, not fully grown* models **do**
      extrapolate/find compatible points
      if no new compatible points
         the model is *fully grown*
      else
         fit a model
         if (GOF ≠ OK)
            reject the last included points
            the model is *fully grown*
         end if

      end if
   end for

   perform selection among all *active* models for the current optimal description
      (only the selected models remain *active*)

end do

output: description of the image in terms of parametric models

---

combination of these two procedures are discussed elsewhere [9]. By properly balancing the trade-offs a computationally efficient and reliable algorithm is obtained which has the feature of growing mostly well-behaved models (in terms of convergence, error, number of compatible points) while at the same time lowering the computation time and space complexity of the procedure. The complete algorithm of the method is given in Table 1.

3.4. Features of the Method

Our approach can be characterized as an approximation of the robust estimation embedded in the framework of recovery and selection. The main features of the method are a high degree of resistance to outliers and the insensitivity to incorrect initial estimates. In this subsection we also briefly discuss the potential of the method to successfully deal with the problem of nonlinear parameter estimation and point out some of the issues regarding the computational complexity of the method.

To maintain the consistency of the data set that belongs to a model throughout the model-recovery process, it is important that the performance of the fitting is constantly monitored. In our case, the procedure dynamically analyzes data consistency enabling the rejection of outliers.

Since the procedure is not limited to prespecified windows we can properly classify the points even if their number is relatively small in comparison with the number of surrounding data points that belong to other models. Thus, the estimate is computed over the whole support of the model which makes the estimates more reliable in the statistical sense.

The idea to initiate a redundant set of models in the image not only enables us to cope with inherently unreliable initial estimates but it also makes possible to successfully deal with *nonlinear* parameter estimation procedures where the convergence to a unique solution is not guaranteed. Starting to develop a redundant set of models from multiple seeds means potentially different initial estimates from which the solution is sought. The sensitivity of the method to the models that get stuck in local minima is thus reduced since such models get rejected during the model-selection procedure if there are other models that better describe the same data.

During model growing, the points that are added to the model are weighted (classified) according to the distance from the extrapolated values of the model. Parameter estimation is then performed without re-weighting of the points in the data set. This is computationally efficient but valid only if the extrapolated values are close to those of the true model. Since a redundant set of models is initiated in the image we can assume that at least some of the models extrapolate well outside their current domains. This approach, of course, presents an approximate solution, in the sense that we can not claim that all the data points that belong to a model will at the end of the procedure (due to the multiple re-estimation of the parameter values) satisfy the compatibility constraint. However, extensive experimental testing has shown that the models selected in the *final result* contain only a negligible number of data points whose deviation exceeds the compatibility constraint, and that those points have almost no influence on the final estimate. This outcome is expected since although we cannot avoid some initial models that are in error, the models that are correct accumulate less error and accept more data points and are thus selected in the final result.

To cope with the computational complexity of the overall method we employed three simplifications, namely in the design of the objective function we consider only pairwise overlaps of the models, we use the greedy algorithm for solving the quadratic Boolean problem, and we iteratively combine the model-recovery and the model-selection procedures. In [9] it is discussed how these simplifications affect the trade-off between the computational efficiency and the reliability of the results. In general, good (optimal) results are obtained as long as the input is well-behaved in the sense that there is a close correspondence between the models and the underlying data.

The proposed method is both general and flexible. Notice that the model-recovery procedure provides the model-selection procedure with the same type of input data regardless of the specific models and parameter estimation techniques (linear, nonlinear). As a consequence, individual procedures for parameter estimation can be replaced with better and more efficient ones when they become available without affecting other modules. In the next section we demonstrate how the method can be applied to various domains, using different models.

4. Recover-and-Select applied to ...

We have tested the proposed method on various domains using different parametric models. Here we present some results of recovering variable-order bivariate polynomials and superquadric models in range images, and parametric curve models in edge images. Since the main idea of the paper is to present a general framework for accurate and robust extraction of parametric models of different types and its relation to some other robust estimation methods, we only briefly mention some of the details pertaining to the particular choice of parametric models and refer the reader to the specific papers.

4.1. ... Surface Models in Range Images

The Recover-and-Select paradigm has been applied to the problem of estimating variable-order bivariate polynomials that are linearly parameterizable in the Euclidean space. We have limited the set of models to planar and second-order surfaces. The surfaces are modeled as functions in the explicit form $z = f(x, y)$. The fitting procedure, which minimizes the sum of squared distances between the data (inliers) and the model in z-direction, yields a system of linear equations for the unknown parameters of the model. The fitting procedure is computationally efficient since the time complexity for updating the parameters of the model is linearly proportional to the number of data points that are added to the model in each iteration.

The decisions regarding data point classification, model acceptance, and model switching (from planar to curved surfaces) are based on simple well-defined thresholds which are all related to the estimated variance of the well-behaved noise. Several experiments have been carried out to test the noise properties, and the stability of the algorithm with respect to perturbations of the thresholds. The interested reader is referred to [11].

Figure 3a shows a noisy range image of a cube[3] (taken from [1]). The object consists of planar and curved surfaces. 143 models were initiated in the image (the

---

[3] The image was provided by Dr. Besl from the University of Michigan.

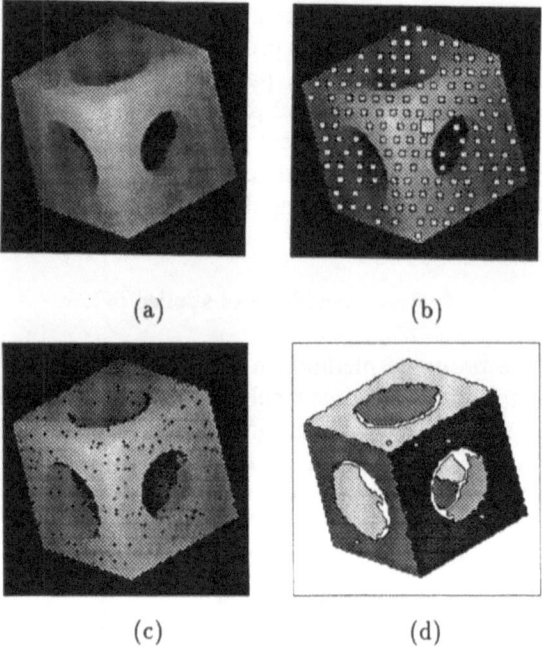

Figure 3. **a** Original range image, **b** seed image, **c** reconstructed image, **d** segmented image

seeds are depicted as white squares in Fig. 3b). Both planar and curved surfaces are reliably recovered. The data points that were detected as outliers are marked in black (Fig. 3c). The domains of the individual models are depicted in Fig. 3d.

4.2. ... Superellipsoids in Range Images

We also used the proposed method to *directly* recover superellipsoids, a subset of superquadric models, from unsegmented range images [12]. This approach is in contrast with standard methods which attempt the recovery of volumetric models only after the data has been presegmented using extensive preprocessing. A super-ellipsoid in general position is defined by an implicit function which involves 11 parameters describing superellipsoid's size, shape, orientation, and position. The parameters of the model are estimated by a nonlinear iterative algorithm [20] which minimizes the sum of squares of the algebraic distances between the data points (classified as belonging to the model) and the superellipsoid, subject to an additional constraint of minimal volume. To decide on the acceptance of the model we calculate an error measure which approximates the Euclidean distance, since the error based on the algebraic distance is inappropriate to evaluate the goodness-of-fit. The search for compatible data points is performed in the neighborhood of the recovered models and is also based on the approximation to the Euclidean distance. Experiments have shown that the method is not sensitive to unreliable

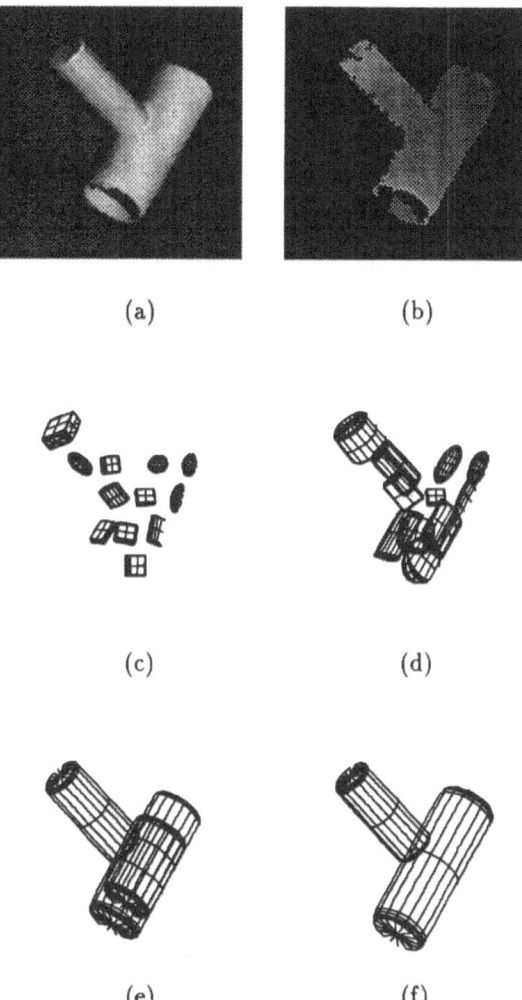

Figure 4. **a** Intensity image, **b** range image, **c** initial models, **d**, **e** selected models after the first and the sixth iteration, respectively, of the recover-and-select algorithm, **f** final result

initial estimates as well as to the models that get stuck in local minima due to the nonlinear nature of the problem.

Figures 4a and 4b show an intensity image and a range image, respectively[4], of a tube and a cylinder that intersect. Twelve models were initiated on the range image at the start of processing (Fig. 4c). Figures 4d and 4e show the partially recovered models during various iterations of the recover-and-select algorithm. The final representation (shown in Fig. 4f) is most compact, consisting of two superellipsoids.

---

[4] The images were provided by Marjan Trobina from the ETH, Switzerland.

4.3. ... *Curve Models in Edge Images*

The method has been applied to the problem of recovering simple geometric curvilinear structures, i.e., straight lines, parabolas, and ellipses in edge images [10]. The fitting procedure, which minimizes the sum of squared Euclidean distances between the data (inliers) and the model, yields a set of nonlinear equations for the unknown parameters of the model. The main advantage of using the Euclidean distance measure (as opposed to, for example, algebraic distance) is that the recovered models extrapolate accurately in the vicinity of the end-points which is necessary for a reliable search for additional consistent data points. Model acceptance and data point classification are also based on the Euclidean distance measure. Initial models are straight line segments and a more complex curve is used only after the goodness-of-fit indicates the inability of the current model to fit the data. Experiments have shown that the procedure is robust with respect to noise (minor edge elements scattered in images, edge elements caused by multiple responses of the edge detector around certain types of edges, etc.).

Figures 5a and 5b show an intensity image and the corresponding edge image, respectively[5]. In Fig. 5c we show the initial models, i.e., small linear segments found

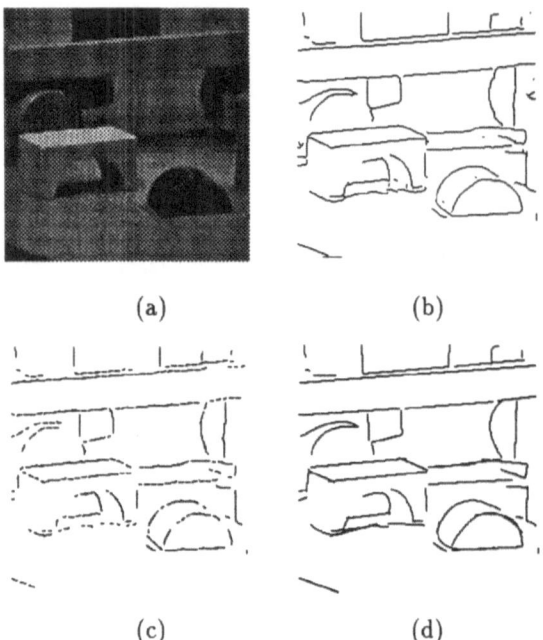

(a)   (b)

(c)   (d)

Figure 5. a Intensity image, **b** edge image, **c** initial models, **d** reconstructed image

---

[5] These two images were kindly supplied by Dr. Etemadi from the University of Surrey.

to be locally consistent in small windows overlaid on the image in a grid-like pattern. Some of the seeds along the parallel lines are missing due to the grid placement. However, a missing seed does not pose a serious problem as long as there are other seeds from which a complete curve can properly be recovered. The final result is depicted in Fig. 5d. Since in this particular case we limited the models to straight lines and parabolas only, one of the circular arcs has a piecewise description consisting of two parabolas.

5. Conclusions and Work in Progress

In this paper we presented a robust method for estimation of parametric models. The main conclusion that we can draw from our work is that reliable results can only be achieved by considering many competitive solutions which are developed independently and selecting those that produce a globally consistent, simple description. The key idea to *independently* and *redundantly* recover parametric models makes the scheme fully parallelizable and thus suitable for implementation on parallel computer architectures[6].

We have also demonstrated that a robust recovery of parametric models can be ensured by iteratively combining data classification and parameter estimation, emphasizing that these two problems are not separable.

The results of the method were presented on two domains (range images, edge images) using different models (surfaces, superquadric models, curve models). However we believe that the proposed method is a general tool that will prove useful for many other tasks of computer vision.

Future work will be directed towards extending the method to other domains using different types of models. Besides, to achieve robustness with respect to the choice of models we plan to independently and simultaneously recover different representations in the image, e.g. surface models and volumetric models, and then use the selection procedure to determine which models are better suited to describe different parts of the image.

Acknowledgement

The author would like to thank Jasna Maver for helpful comments.

References

[1] Besl, P. J.: Surfaces in range image understanding. New York: Springer 1988.
[2] Besl, P. J., Birch, J. B., Watson, L. T.: Robust window operators. In: Proceedings of the 2nd International Conference on Computer Vision, pp. 591–600. Tampa, FL, December 1988. Washington: IEEE Computer Society Press 1988.

---

[6] The procedure for variable-order bivariate polynomials (surfaces) has been implemented on the Connection Machine.

[3] Chen, D. S.: A data-driven intermediate level feature extraction algorithm. IEEE Trans. Pattern Anal. Machine Intell. *PAMI-11*, 749–758 (1989).

[4] Fischler, M. A., Bolles, R. C.: Random sample consensus: A paradigm for model fitting with applications to image analysis and automated cartography. Commun. ACM *24*, 381–395 (1981).

[5] Fua, P., Hanson, A. J.: Objective functions for feature discrimination. In: Proceedings of the 11th International Joint Conference on Artificial Intelligence, pp. 1596–1602. Detroit, MI, August 1989. San Mateo: Morgan Kaufmann.

[6] Huber, P. J.: Robust statistics. New York: Wiley 1981.

[7] Kim, D. Y., Kim, J. J., Meer, P., Mintz, D., Rosenfeld, A.: Robust computer vision: A least-median of squares based approach. In: Proceedings of the Image Understanding Workshop, pp. 1117–1134. Palo Alto, CA, May 1989. San Mateo: Morgan Kaufmann.

[8] Leclerc, Y. G.: Constructing simple stable descriptions for image partitioning. Int. J. Comput. Vision *3*, 73–102 (1989).

[9] Leonardis, A.: Image analysis using parametric models: Model-recovery and model-selection paradigm. PhD thesis, Faculty of Electrical Engineering and Computer Science, University of Ljubljana, Slovenia, May 1993. Technical Report LRV-93-3.

[10] Leonardis, A., Bajcsy, R.: Finding parametric curves in an image. In: Proceedings of The Second European Conference on Computer Vision—ECCV-92, Santa Margherita Ligure, Italy (Sandini, G., ed.), pp. 653–657. Berlin, Heidelberg, New York, Tokyo: Springer 1992 (Lecture Notes in Computer Science, Vol. 588).

[11] Leonardis, A., Gupta, A., Bajcsy, R.: Segmentation of range images as the search for geometric parametric models. Int. J. Comput. Vision *14*, 253–277 (1995).

[12] Leonardis, A., Solina, F., Macerl, A.: A direct recovery of superquadric models in range images using recover-and-select paradigm. In: Proceedings of The Third European Conference on Computer Vision—ECCV-94, Stockholm, Sweden (Eklundh, J.-O., ed.), pp. 309–318. Berlin, Heidelberg, New York, Tokyo: Springer 1994 (Lecture Notes in Computer Science, Vol. 800).

[13] Li, G.: Robust regression. In: Exploring data tables, trends and shapes (Hoaglin, D. C., Mosteller, F., Tukey, J. W., eds.), pp. 281–343. New York: J. Wiley 1985.

[14] Meer, P., Mintz, D., Rosenfeld, A., Kim, D. Y.: Robust regression methods for computer vision: A review. Int. J. Comput. Vision *6*, 59–70 (1991).

[15] Mirza, M. J., Boyer, K. L.: Performance evaluation of a class of M-estimators for surface parameter estimation in noisy range data. IEEE Trans. Robotics Automation *9*, 75–85 (1993).

[16] Pentland, A. P.: Part segmentation for object recognition. Neural Comput. *1*, 82–91 (1989).

[17] Rousseuw, P. J., Leroy, A. M.: Robust regression and outlier detection. New York: Wiley 1987.

[18] Schunck, B. G.: Robust computational vision. In: Proceedings of the International Workshop on Robust Computer Vision, Seattle, WA, October 1990.

[19] Sinha, S. S., Schunck, B. G.: A two-stage algorithm for discontinuity-preserving surface reconstruction. IEEE Trans. Pattern Anal. Machine Intell. *PAMI-14*, 36–55 (1992).

[20] Solina, F., Bajcsy, R.: Recovery of parametric models from range images: The case for superquadrics with global deformations. IEEE Trans. Pattern Anal. Machine Intell. *PAMI-12*, 131–147 (1990).

Dr. A. Leonardis
Computer Vision Laboratory
Faculty of Electrical Engineering
and Computer Science
University of Ljubljana
e-mail: Ales.Leonardis@fer.uni-lj.si

Computing Suppl. 11, 131–148 (1996)

Computer Vision and Mathematical Morphology*

J. B. T. M. Roerdink, Groningen

Abstract

Computer Vision and Mathematical Morphology. Mathematical morphology as originally developed by Matheron and Serra is a theory of set mappings, modeling binary image transformations, which are invariant under the group of Euclidean translations. This framework turns out to be too restricted for many applications, in particular for computer vision where group theoretical considerations such as behavior under perspective transformations and invariant object recognition play an essential role. So far, symmetry properties have been incorporated by assuming that the allowed image transformations are invariant under a certain commutative group. This can be generalized by dropping the assumption that the invariance group is commutative. To this end we consider an arbitrary homogeneous space (the plane with the Euclidean translation group is one example, the sphere with the rotation group another), i.e. a set $\mathscr{X}$ on which a transitive but not necessarily commutative transformation group Γ is defined. As our object space we then take the Boolean algebra $\mathscr{P}(\mathscr{X})$ of all subsets of this homogeneous space. Generalizations of dilations, erosions, openings and closings are defined and several representation theorems can be proved. We outline some of the limitations of mathematical morphology in its present form for computer vision and discuss the relevance of the generalizations discussed here.

Key words: Mathematical morphology, image processing, transitive group action, non-commutative transformation group, Minkowski operations, invariance, computer vision.

1. Introduction

Mathematical morphology was originally developed at the Paris School of Mines as a set-theoretical approach to image analysis [10, 21]. It has a strong algebraic component, studying image transformations with a simple geometrical interpretation and their decomposition and synthesis in terms of set operations. Although the main object of our present study is the algebraic approach we emphasize that our primary motivation comes from the geometrical side, in the sense that various image transformations used in mathematical morphology today (dilations, erosions, openings, closings) have a straightforward geometrical analogue in a more general context. It is then a natural question to ask whether a corresponding algebraic description can be found.

---

* Presented at the Schloß Dagstuhl Seminar on "Theoretical Foundations of Computer Vision", Schloß Dagstuhl, Germany, March 13–18, 1994.

In the original approach a two-dimensional image of, let us say, a planar section of a porous material is modeled as a subset X of the plane. In order to reveal the structure of the material, the image is probed by translating small subsets B, called *structuring elements*, of various forms and sizes over the image plane and recording the locations h where certain relations (e.g. 'B_h included in X', 'B_h hits X', etc.) between the image X and the translate B_h of the structuring element B over the vector h are satisfied. In this way one can construct a large class of image transformations which are invariant under the Euclidean translation group. The underlying idea here is that the form or shape of objects in the image does not depend on the relative location with respect to an arbitrary origin and that therefore the transformations performed on the image should respect this. Notice that the basic object of study, the 'object space', is not the reference space (the plane in our example) itself, but the collection of subsets of this reference space, and the transformations defined on this collection of subsets.

In practice one encounters various situations where this framework is too restrictive. Certain images show a clear radial symmetry with an intrinsic origin. In this case we need image transformations which are adapted to the symmetries of this polar structure. Now one obtains a straightforward generalization of Euclidean morphology by replacing the Euclidean translations by an arbitrary *abelian* (*commutative*) group [5, 18]. In the case of the example mentioned above, this would be the group generated by rotations and multiplications with respect to the origin. Here the size of the structuring element increases with increasing distance from the origin. Another example occurs in the analysis of traffic scenes, where the goal is to recognize the shape of automobiles with a camera on a bridge overlooking a highway [2]. In this case the size of the structuring element has to be adapted according to the law of perspective. In Section 3 we will show that in this case there is again invariance under a commutative group. Notice that in the two examples just mentioned we have a variable structuring element as a function of position. In fact we will argue that without a concept of invariance (under a group, or otherwise), one cannot even give a meaningful answer to the question when sets at different locations are 'of the same shape' or not.

Instead of changing the symmetry group of the object space one may generalize the object space itself to complete lattices. This is the approach initiated by Serra and Matheron [21, 22], as well as Heijmans [5]. A general study of this topic has recently been made by Heijmans and Ronse, see [6, 7, 20]. A lattice formulation also is in order for studying grey-level images.

One may drop the assumption that the invariance group is commutative [15]. To this end we consider an arbitrary *homogeneous space*, i.e. a set $\mathscr{X}$ on which a transitive but not necessarily commutative group Γ of invertible transformations is defined. Here *transitive* means that for any pair of points in the set there is a transformation in the group which maps one point on the other. If this mapping is unique we say that the transformation group is *simply transitive* or *regular*. As the object space of interest from a morphological point of view we take here the Boolean algebra of all subsets of this homogeneous space.

We present some examples for basic motivation. First of all one may extend Euclidean morphology in the plane by including rotations. This case has been extensively discussed in [13]. In that case it is appropriate to use the full Euclidean group of motions (the group generated by translations and rotations) as (non-commutative) invariance group. This is for example the basic assumption made in integral geometry to give a complete characterization (Hadwiger's Theorem) of functionals of compact, convex sets in $\mathbb{R}^n$ [4]. As our second basic example we mention the sphere with its symmetry group of three-dimensional rotations, again a non-abelian group. A detailed investigation of this case is presented in [14]. A third example is the area of path planning or motion planning. Here the problem is to find a path for an object, say a robot or a car, moving in a space (called 'work space') with obstacles [16].

In the area of computer vision, mathematical morphology has so far not made a breakthrough. This is the case for several reasons. First, there is the limitation that mathematical morphology considers a 2D image as a direct representation of the structure which is analyzed. In particular the question of how to take the projective geometry of the imaging process into account has not been discussed. This is a serious shortcoming, since clearly the symmetry of a two-dimensional plane is not the same as the symmetry of the three-dimensional world of which it is a projection. It is here that we try to make some progress by the methods discussed in this paper.

Another, even more serious problem, which will not be addressed here, is that of *occlusion* of objects in 3D. The basic ingredients of mathematical morphology are ordering of sets by inclusion and transformations adapted to this ordering. So far it is not clear at all how mathematical morphology can be modified to deal with this problem and related ones, such as influence of lightning conditions, surface properties, etc. This situation was pregnantly summarized by Ronse [19] in the aphorism: *Mathematical morphology is flat.*

A first attempt towards morphological analysis of patterns on curved surfaces has been made in [17] using techniques from differential geometry. It remains to be seen to what extent this will become a useful tool for computer vision too. For general questions of invariance in computer vision, see for example Mundy et al. [11].

The organization of the paper is as follows. First we introduce the general framework of mathematical morphology on Euclidean space, followed by its generalization to arbitrary homogeneous spaces (Section 2). Then the applications to computer vision are studied. In Section 3 we take the projective geometry of the imaging process into account and explain how to define morphological operations on images of a planar patch produced by perspective projection, where the allowed 3D motion is restricted to translation within the object plane. Finally in Section 4 a problem from robot vision for path planning is approached by morphological methods.

2. Preliminaries

In this section we first outline some elementary concepts and results from classical
Euclidean morphology. Then we generalize this to homogeneous spaces.

2.1. Euclidean Morphology

Consider the space E, where $E = \mathbb{R}^n$ or $E = \mathbb{Z}^n$. Denote by $\mathscr{P}(E)$ the power set (set
of all subsets) of E. The classical Minkowski addition and subtraction for subsets
X, A of E are given by

$$X \oplus A = \bigcup_{a \in A} X_a, \tag{1}$$

$$X \ominus A = \bigcap_{a \in A} X_{-a}, \tag{2}$$

where

$$X_a = \tau_a(X) = \{x + a \colon x \in X\},$$

is the translate of X over the vector $a \in E$, $x + y$ is the sum of x and y, and $-x$ the
reflection of x. It can be shown that

$$X \oplus A = \{h \in E \colon \check{A}_h \ X\}, \tag{3}$$

where $\check{A} = \{-a \colon a \in A\}$ is the *reflection* of A and $A \ B$ (A 'hits' B) is a general
notation for $A \cap B \neq \varnothing$.

Two characteristic properties of dilation are:

$$\textit{Distributivity w.r.t. union:} \quad \left(\bigcup_{i \in I} X_i \right) \oplus A = \bigcup_{i \in U} (X_i \oplus A) \tag{4}$$

$$\textit{Translation invariance:} \quad (X \oplus A)_h = X_h \oplus A. \tag{5}$$

Similar properties hold for the erosion with intersection instead of union. A consequence of the distributivity property is that dilation and erosion are *increasing*
mappings. (A mapping ψ is called increasing when for all X, $Y \in \mathscr{P}(E)$, $X \subseteq Y$
implies that $\psi(X) \subseteq \psi(Y)$.)

Other important increasing transformations are the opening and closing by a
structuring element A:

$$\textit{Opening:} \quad X \circ A := (X \ominus A) \oplus A = \bigcup_{h \in E} \{A_h \colon A_h \subseteq X\} \tag{6}$$

$$\textit{Closing:} \quad X \bullet A := (X \oplus A) \ominus A = \bigcap_{h \in E} \{(\check{A}^c)_h \colon (\check{A}^c)_h \supseteq X\}. \tag{7}$$

The opening is the union of all the translates of the structuring element which
are included in the set X. Opening and closing are related by Boolean duality:

$(X^c \circ A)^c = X \bullet \check{A}$. A more general definition of dilations, erosions, openings and closings can be given in the framework of complete lattices [7].

2.2. Generalized Minkowski Operators

On any group Γ one can define generalizations of the Minkowski operations [15]. Recall that a *dilation* (*erosion*) is a mapping commuting with unions (intersections). For any subsets G, H of Γ define the dilation

$$\delta(G) := G \overset{\Gamma}{\oplus} H := \bigcup_{h \in H} Gh = \bigcup_{g \in G} gH, \tag{8}$$

which generalizes the Minkowski addition to non-commutative groups. Here

$$gH := \{gh: h \in H\}, \qquad Gh := \{gh: g \in G\},$$

with gh the group product of g and h. Similarly, define the erosion (h^{-1} is the group inverse of h)

$$\varepsilon(G) := G \overset{\lambda}{\ominus} H := \bigcap_{h \in H} Gh^{-1},$$

which generalizes the Minkowski subtraction. Both mappings are *left-invariant*, e.g.

$$\delta(gG) = g\delta(G), \qquad \forall g \in \Gamma.$$

This is the reason for the superscript 'λ' on the '$\ominus$' symbol. For later use we also define the *inverted* set G^{-1} of G by

$$G^{-1} = \{g^{-1}: g \in G\}. \tag{9}$$

Duality by complementation is expressed by the formula $(G \overset{\Gamma}{\oplus} H)^c = G^c \overset{\lambda}{\ominus} H^{-1}$.

2.3. Group Actions

Let $\mathscr{X}$ be a non-empty set, Γ a transformation group on $\mathscr{X}$, that is, each element $g \in \Gamma$ is a mapping $g: \mathscr{X} \to \mathscr{X}$, satisfying

(i) $gh(x) = g(h(x))$
(ii) $e(x) = x$,

where e is the unit element of Γ, and gh denotes the product of two group elements g and h. The *inverse* of an element $g \in \Gamma$ will be denoted by g^{-1}. Instead of $g(x)$ we will also write gx. We say that Γ is a *group action* on $\mathscr{X}$ [1, 12, 23].

The group Γ is called *transitive on* $\mathscr{X}$ if for each x, $y \in \mathscr{X}$ there is a $g \in \Gamma$ such that $gx = y$, and *simply transitive* when this element g is unique. A *homogeneous space* is a pair $(\Gamma, \mathscr{X})$ where Γ is a group acting transitively on $\mathscr{X}$. Any transitive abelian permutation group Γ is simply transitive. If Γ acts on $\mathscr{X}$, the *stabilizer* of $x \in \mathscr{X}$ is the subgroup $\Gamma_x := \{g \in \Gamma: gx = x\}$. Let ω be an arbitrary but fixed point of $\mathscr{X}$,

	simply transitive	multi-transitive
commutative	Euclidean morphology	
non-commutative	group morphology	group action morphology

Figure 1. Classification of transformation groups and the associated morphologies

henceforth called the *origin*. The stabilizer Γ_ω will be denoted by Σ from now on:

$$\Sigma := \Gamma_\omega = \{g \in \Gamma: g\omega = \omega\}. \tag{10}$$

The set of group elements which map ω to a given point x is called a *left coset* and denoted by

$$g_x\Sigma := \{g_x s: s \in \Sigma\}. \tag{11}$$

Here g_x is a representative (an arbitrary element) of this coset.

In Fig. 1 we give a classification of transformation groups and the associated morphologies.

2.4. Examples

In the following we present two examples. In each case Γ denotes the group and $\mathscr{X}$ the corresponding set.

Example 1. $\mathscr{X} =$ Euclidean space $\mathbb{R}^n$, $\Gamma =$ the Euclidean translation group **T**. **T** is abelian. Elements of **T** can be parametrized by vectors $h \in \mathbb{R}^n$, with τ_h the translation over the vector h:

$$\tau_h x = x + h, \qquad h, x \in \mathbb{R}^n. \tag{12}$$

Example 2. $\mathscr{X} =$ Euclidean space $\mathbb{R}^n$ $(n = 2)$, $\Gamma =$ the Euclidean motion group $\mathbf{M} := E^+(3)$ (proper Euclidean group, group of rigid motions), i.e. the group generated by translations and rotations (see [13]). The subgroup leaving a point p fixed is the set of all rotations around that point. **M** is not abelian. The collection of translations forms the Euclidean translation group **T**. The stabilizer, denoted by **R**, equals the circle group S^1 (also commutative) of rotations around the origin. Let τ_h denote the translation over the vector $h \in \mathbb{R}^2$ and ρ_ϕ^p the rotation over an angle ϕ around the point p. The following relations, whose proof is left to the reader, are needed in the sequel:

$$\rho_\phi^p = \tau_p \rho_\phi^0 \tau_{-p} = \tau_{p - \rho_\phi^0 p} \rho_\phi^0, \tag{13}$$

$$\rho_\phi^p \tau_h = \tau_{\rho_\phi^0 h} \rho_\phi^p. \tag{14}$$

Let $\gamma_{h,\phi}$ denote a rotation around the origin followed by a translation:

$$\gamma_{h,\phi} = \tau_h \rho_\phi^0, \qquad h \in \mathbb{R}^2, \phi \in S^1. \tag{15}$$

Any element of **M** can be written in this form. Using the rules (13)–(14) one finds the multiplication rule

$$\gamma_{h,\,\phi}\gamma_{h',\,\phi'} = \gamma_{h+\rho_\phi^0 h',\,\phi+\phi'}.\tag{16}$$

Geometrical representation: The following representation is useful in this case [13, 15]. Attach a set of unit vectors $\vec{v}$ with direction varying over the unit circle to each point in the plane. We call $\mathbf{p} := (x,\vec{v})$ a *pointer.* Given any pointer $\mathbf{p} = (x,\vec{v})$, there is a unique element of the group Γ which maps a fixed pointer $\mathbf{b} := (\omega,\vec{e}_1)$, called *the base pointer* to $\mathbf{p}$, where ω is the origin and $\vec{e}_1$ a unit vector in the x-direction. So the pointer $\mathbf{p} = (x,\vec{v})$, where $\vec{v} = (\cos\phi,\sin\phi)$, represents the motion $\gamma_{x,\,\phi}$. In this representation, the rotation group **R** is the set of unit vectors attached to the origin and **T** is represented by the collection of horizontal unit vectors attached to points of $\mathbb{R}^2$. In the discrete case we will use a hexagonal grid and replace **R** by a finite group **H** consisting of rotations over multiples of 60°. The coset $\tau_x\mathbf{R}$ is represented on the hexagonal grid by the six unit vectors attached to the point x [15]. An example is given in Fig. 2, cf. also Fig. 6 below.

2.5. Morphological Operations on Homogeneous Spaces

One can construct morphological operations on an arbitrary homogeneous space $\mathscr{X}$ as follows. Define the 'origin' ω to be an arbitrarily chosen point or $\mathscr{X}$. To each subset X of $\mathscr{X}$ we associate all elements of the group which map the origin ω to an element of X. We also can go back from the group Γ to the space $\mathscr{X}$ by associating to each subset G of Γ the collection of all points $g\omega$ where g ranges over G. This is summarized in the following definition. As above, the symbol $\mathscr{P}(A)$ denotes the set of all subsets of A, A an arbitrary set.

Definition 3. *The lift* $\vartheta\colon \mathscr{P}(\mathscr{X}) \to \mathscr{P}(\Gamma)$ *and canonical projection* $\pi\colon \mathscr{P}(\Gamma) \to \mathscr{P}(\mathscr{X})$ *are defined by*

$$\vartheta(X) = \{g \in \Gamma\colon g\omega \in X\}, \qquad X \subseteq \mathscr{X}$$
$$\pi(G) = \{g\omega\colon g \in G\}, \qquad G \subseteq \Gamma.$$

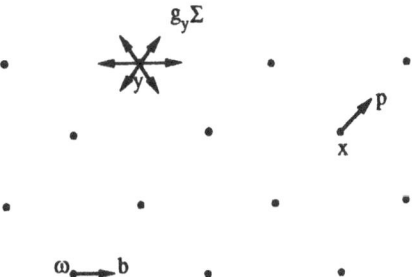

Figure 2. Representation of the Euclidean motion group. *b* Base pointer. *p* Pointer with base point *x*. $g_y\Sigma$ Collection of group elements which map the origin ω to *y*. Each pointer represents a unique group element

For the case of Example 2, these formulas specialize to

$$\vartheta(X) = \bigcup_{x \in X} \tau_x \mathbf{R} = \tau(X) \overset{M}{\oplus} \mathbf{R}, \tag{17}$$

where

$$\tau(X) := \{\tau_x : x \in X\}. \tag{18}$$

In [13, 15] a construction was performed of various morphological operators between the distinct lattices $\mathscr{P}(\mathscr{X})$ and $\mathscr{P}(\Gamma)$. Here we restrict ourselves to dilations. That is, consider the mapping—referred to as the *hitting function* below—which associates to a subset X of $\mathscr{X}$ the set of group elements $g \in \Gamma$ for which the translated set $gB := \{gb : b \in B\}$ hits X (cf. (3)):

$$\mathscr{D}_B(X) := \{g \in \Gamma : gB \; X\}. \tag{19}$$

Then it was shown in [15] that

$$\mathscr{D}_B(X) = \{g \in \Gamma : g\vartheta(B) \; \vartheta(X)\}$$
$$= \vartheta(X) \overset{\Gamma}{\oplus} \vartheta^{-1}(B). \tag{20}$$

Here $\vartheta(X) \overset{\Gamma}{\oplus} \vartheta^{-1}(B)$ is a generalized Minkowski operation on subsets of Γ as defined in (8), with $\vartheta^{-1}(B)$ the inverted set of $\vartheta(B)$, cf. (9). This mapping is

- a dilation $\mathscr{P}(\mathscr{X}) \to \mathscr{P}(\Gamma)$;

- invariant under Γ, that is

$$\mathscr{D}_B(g(X)) = g\mathscr{D}_B(X), \qquad g \in \Gamma.$$

More generally, if A is a subset of Γ and $x \mapsto g_x$ a Γ-invariant function from $\mathscr{X}$ to Γ, the mapping

$$\delta_A^{\Gamma}(X) := \bigcup_{x \in X} g_x A, \tag{21}$$

is a dilation $\mathscr{P}(\mathscr{X}) \to \mathscr{P}(\Gamma)$ which is Γ-invariant. For this reason we speak sometimes of *group dilations*, or Γ-*dilations*. Dilations (and erosions) from $\mathscr{P}(\Gamma)$ to $\mathscr{P}(\mathscr{X})$ or from $\mathscr{P}(\mathscr{X})$ to $\mathscr{P}(\mathscr{X})$ can be constructed similarly but are not needed in the sequel.

2.6. The Role of Symmetry Groups in Shape Description

We make a short remark concerning the problem how to define 'shape', which is known to present great difficulties and is a recurring theme in the image processing and computer vision literature. Often, 'shape' is defined as referring to those properties of geometrical figures which are invariant under the Euclidean similarity group [8]. Intuitively spoken, one first has to bring figures to a standard location, orientation and scale before being able to 'compare' them. Now it is not necessary to restrict oneself to the similarity group, although in the absence of any form of group invariance there is no way at all for comparing figures. In the present context the following definition seems appropriate.

Definition 4. *Let $\mathscr{X}$ be a set, Γ a group acting on $\mathscr{X}$. Two subsets X, Y of $\mathscr{X}$ are said to have* the same shape *with respect to Γ, or the same Γ-shape, if they are Γ-equivalent, meaning that there is a $g \in \Gamma$ such that $Y = gX$. If no such $g \in \Gamma$ exists, X and Y are said to have* different Γ-shape.

In essence this definition goes back to F. Klein's Erlanger Program (1872), which considers geometry to be the study of transformation groups and the properties invariant under these groups [9]. So in Euclidean morphology, all translates of a set X by the Euclidean translation group **T** have the same **T**-shape. Adding rotations to get the Euclidean motion group **M**, rotated versions of X or its translates have the same **M**-shape as X. Extreme cases are (*i*) $\Gamma = \{id\}$, so that all sets have different shape, and (*ii*) $\Gamma = \mathsf{Sym}_{\mathscr{X}}$ (full permutation group), in which case all sets with the same cardinality have the same shape.

3. Relevance for Computer Vision

Above we have outlined how Euclidean morphology can be generalized to arbitrary homogeneous spaces $(\mathscr{X}, \Gamma)$, where the group Γ acting on $\mathscr{X}$ is not necessarily commutative. The case where Γ acts simply transitively on $\mathscr{X}$ leads to the study of transformations of subsets of an arbitrary group which are invariant under either left or right group translations. The general case where Γ acts transitively on $\mathscr{X}$ has been treated by (*i*) mapping the subsets of $\mathscr{X}$ to subsets of Γ; (*ii*) using the result for the simply transitive case, and (*iii*) projecting back to the original space. The main result is that the scope of mathematical morphology is widened to situations where a non-commutative group is involved.

3.1. Perspective Transformations

The application to computer vision concerns in particular the behavior of morphological operators under perspective transformations. This is a difficult problem and we confine us here to a sketch of a simplified case which nevertheless exhibits some of the essential ingredients of the general case. The emphasis will be on combining invariance concepts as relevant for computer vision in general with the construction of invariant morphological operators as outlined above. In a real 3D situation we will have to consider the full (non-abelian) group of projective transformations. But this only makes sense if at the same time we take the occlusion problem into account (see the introduction).

Consider the perspective transformation of a plane shape lying in a plane V which we call the object plane, see Fig. 3. As is well known, 3D Euclidean motions of the plane V induce 2D projection transformations on the image plane under perspective projection. Subgroups of the projection group apply when the motion of the planar object is constrained, for example to rotations and/or translations *within* the object plane.

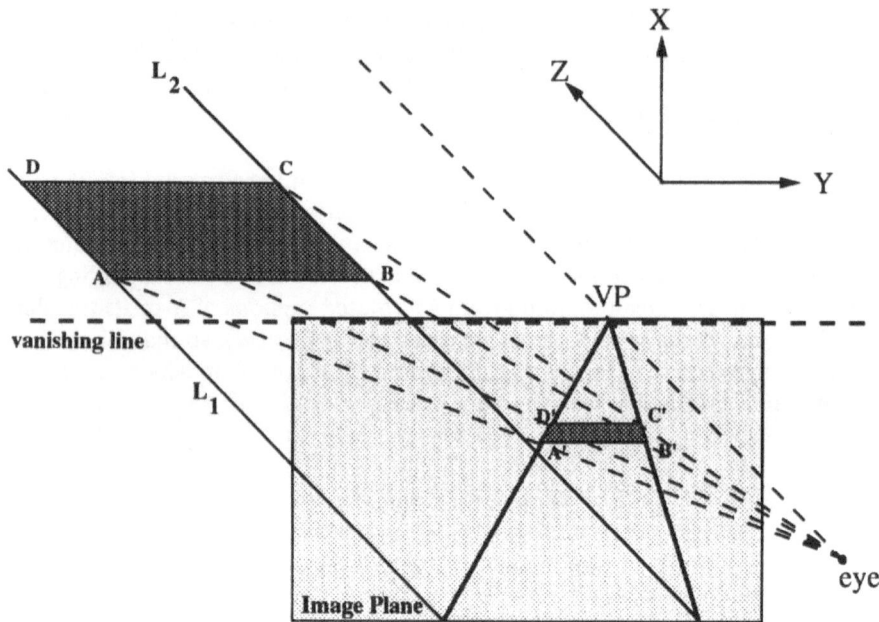

Figure 3. Perspective projection of a planar object *ABCD* on the image plane

Effect of object plane translations: The case to which we confine ourselves here is that of a translation only within the object plane. We will derive the corresponding morphological operators such as dilation and erosion, and show how this reduces to the Euclidean case when the object plane is parallel to the image plane.

Let the object plane V have equation

$$AX + BY + CZ + D = 0. \tag{22}$$

We use a camera-centered coordinate system with coordinates X, Y, Z in object space, so the image plane $\mathscr{X}$ is parallel to the X–Y plane at distance f above it (f is the focal length) with the viewpoint located at the origin $(0,0,0)$. We use coordinates x, y in the image plane, with the x and y axes parallel to the X and Y axes, respectively. We also assume that $C \neq 0$, i.e., the object plane V has a nonzero slope w.r.t. the Z-axis.

Then we have that a point (X, Y, Z) on the object plane is projected to the point (x, y) on the image plane, where

$$x = fX/Z, \qquad y = fY/Z.$$

We will use π to denote this projection:

$$(x, y) = \pi(X, Y, Z). \tag{23}$$

Conversely, a point $\vec{x} = (x, y)$ on the image plane and below the vanishing line is mapped to a unique point $\vartheta(x, y) = (Zx/f, Zy/f, Z)$ on the object plane, where Z is given by (use (22)):

$$Z = \frac{-Df}{Ax + By + Cf}.$$

(24)

Therefore we find

$$\vartheta(x, y) = \frac{-D}{N(\vec{x})}(x, y, f)$$

(25)

where

$$N(\vec{x}) = Ax + By + Cf.$$

(26)

Now consider a translation of the plane shape within the object plane V. The translation vector $\vec{\tau}$ has to be parallel to V, i.e., perpendicular to the normal (A, B, C) of V, so

$$A\tau_1 + B\tau_2 + C\tau_3 = 0.$$

(27)

Using the notation of Section 2, the corresponding transformation $g_{\vec{\tau}}$ acting on points of the image plane is given by

$$g_{\vec{\tau}}\vec{x} = \pi[\vartheta(\vec{x}) + \vec{\tau}].$$

(28)

Using (25) and (23) we obtain

$$g_{\vec{\tau}}\vec{x} = \frac{f}{-Df/N(\vec{x}) + \tau_3}(-Dx/N(\vec{x}) + \tau_1, -Dy/N(\vec{x}) + \tau_2).$$

(29)

It is clear that $g_{\vec{\tau}}g_{\vec{\tau}'} = g_{\vec{\tau}+\vec{\tau}'}$, that $g_{\vec{\tau}}^{-1} = g_{-\vec{\tau}}$ and that $g_{\vec{0}}$ is the identity transformation where $\vec{0}$ denotes the origin of the image plane.

Parametrization of the translations: There is a 1-1 correspondence between translations parallel to V and points of the image plane below the vanishing line: given such a point $\vec{x} = (x, y)$, there is a unique translation $\vec{\tau}(\vec{x})$ such that

$$g_{\vec{\tau}(\vec{x})}\vec{0} = \vec{x}.$$

(30)

So from (29) we see that $\vec{\tau}(\vec{x}) = (\tau_1, \tau_2, \tau_3)$ has to satisfy

$$\frac{f}{-D/C + \tau_3}(\tau_1, \tau_2) = (x, y).$$

(31)

This is a pair of equations which together with (27) has the unique solution

$$\tau_1 = \frac{-Dx}{Ax + By + Cf}$$

(32)

$$\tau_2 = \frac{-Dy}{Ax + By + Cf}$$

(33)

$$\tau_3 = (D/C)\frac{Ax + By}{Ax + By + Cf}$$

(34)

Using (26) we can write this more compactly as

$$\vec{\tau}(\vec{x}) = \frac{D}{N(\vec{x})}\left(-x, -y, \frac{N(\vec{x})}{C} - f\right).\tag{35}$$

Group operation on points of the image plane: Using the parametrization found above we can now define a group operation between points $\vec{x} = (x, y)$ and $\vec{x}' = (x', y')$ of the image plane below the vanishing line, which mimics the action of the translation group **T** on V:

$$\vec{x} \overset{P}{*} \vec{x}' := g_{\vec{\tau}(\vec{x})} g_{\vec{\tau}(\vec{x}')} \vec{0}$$

$$= g_{\vec{\tau}(\vec{x}) + \vec{\tau}(\vec{x}')} \vec{0}\tag{36}$$

Here the superscript 'P' in $\overset{P}{*}$ stands for 'perspective'.

Using (29) and (35) we obtain after doing the algebra

$$\vec{x} \overset{P}{*} \vec{x}' = \frac{fC(x/N(\vec{x}) + x'/N(\vec{x}'), y/N(\vec{x}) + y'/N(\vec{x}'))}{-1 + fC/N(\vec{x}) + fC/N(\vec{x}')}$$

$$= \frac{fC(\vec{x}/N(\vec{x}) + \vec{x}'/N(\vec{x}'))}{-1 + fC/N(\vec{x}) + fC/N(\vec{x}')}.\tag{37}$$

Note that the result does not depend on D: the distance of the object plane V to the camera is irrelevant.

The identity element under this group operation is $\vec{0} = (0,0)$ and the inverse of a point $\vec{x}$ is

$$\vec{x}^{-1} = g_{-\vec{\tau}(\vec{x})} \vec{0}\tag{38}$$

$$= \frac{-Cf}{2N(\vec{x}) - Cf} \vec{x}.\tag{39}$$

Remark 5. Is is clear from (37) that the group operation $\overset{P}{*}$ is abelian. This is of course due to the commutativity of the translation group **T** acting on V.

Remark 6. As a special case we consider an object plane *parallel* to the image plane: $A = B = 0$. In that case we find that

$$\vec{x} \overset{P}{*} \vec{x}' = \vec{x} + \vec{x}'.\tag{40}$$

This is the ordinary vector addition leading to the classical Minkowski addition and subtraction of Section 2.1.

Perspective Minkowski operation: Now that we have found a commutative group operation on points of the image plane, it is time to define the corresponding 'perspective' Minkowski operations using the standard recipe of Section 2.2. We formulate the result as a theorem.

Theorem 7. *Let* $\overset{P}{*}$ *be the group operation between point* $\vec{x} = (x, y)$ *and* $\vec{x}' = (x', y')$ *of the image plane induced by translations* **T** *within the object plane defined by the equation* $AX + BY + CZ + D = 0$ $(C \neq 0)$:

$$\vec{x} \overset{P}{*} \vec{x}' = \frac{fC(\vec{x}/N(\vec{x}) + \vec{x}'/N(\vec{x}'))}{-1 + fC/N(\vec{x}) + fC/N(\vec{x}')},$$

(41)

where $N(\vec{x}) = Ax + By + Cf$.

Then the dilation $\delta(G)$ *and erosion* $\varepsilon(G)$ *of subsets* G *of the image plane* $\mathscr{X}$ *invariant w.r.t. the group* **P** *induced by* **T** *are given by:*

$$\delta(G) := G \overset{P}{\oplus} H := \bigcup_{\vec{x}' \in H} G\vec{x}' = \bigcup_{\vec{x} \in G} \vec{x} H$$

(42)

$$\varepsilon(G) := G \overset{P}{\ominus} H := \bigcap_{\vec{x}' \in H} G\vec{x}'^{-1}$$

(43)

where

$$\vec{x} H := \{\vec{x} \overset{P}{*} \vec{x}' : \vec{x}' \in H\}, \qquad G\vec{x}' := \{\vec{x} \overset{P}{*} \vec{x}' : \vec{x} \in G\}.$$

Practical relevance: Suppose one takes pictures by a camera from a scene, such as occurs in the analysis of traffic scenes. If morphological analysis of such pictures is applied in order to recognize the shape of automobiles, the size of the structuring element has to be adapted according to the law of perspective, see Fig. 4. This is exactly what the theorem encapsulates.

4. Robot Vision for Path Planning

Here the problem is to find a path for an object, say a robot or a car, moving in a space with obstacles. The problem falls apart into two distinct subproblems [3]. First, the *empty-space problem*: find the allowed states of the robot. Any possible configuration of the robot is represented as a point in a *configuration space* $\mathscr{C}$, whose dimensionality equals the number of degrees of freedom of the robot. Points in $\mathscr{C}$ such that a robot in that configuration would collide with any of the obstacles

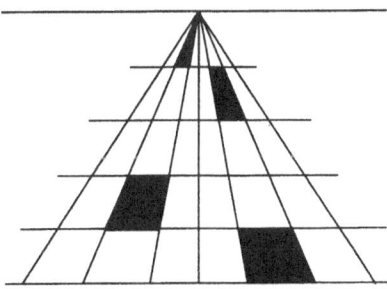

Figure 4. Adapting the structuring element (black regions) under perspective projection

in work space are 'forbidden'. The set of allowed points of $\mathscr{C}$ is called 'empty-space'. The second problem to be solved is the *find-path problem*: find a trajectory in the empty space, where the definition of 'trajectory' has to specify which transitions between allowed states are permissible. An approach based on mathematical morphology is able to solve the empty-space problem. If only translations of the robot are possible, one can find allowed positions of the (arbitrarily chosen) center of the robot by an ordinary erosion of the space outside the obstacles, where the structuring element B is the robot itself. Equivalently, one may perform the dilation by the reflected set $\check{B}$ of the set of obstacles to find the *forbidden* positions of the center of the robot. If the robot has rotational degrees of freedom one has to perform dilations with all rotated versions of the robot. The problem becomes even more difficult when the robot has internal degrees of freedom, see for example, for a robot with several rotating joints. For a full discussion see [16]. In this subsection the allowed motion group equals either the translation group or the translation-rotation group.

4.1. Translations Only

Consider a robot moving in the plane $\mathbb{R}^2$ with obstacles. The robot corresponds to a subset B of the plane and the obstacles to another subset, say X. The problem is to find the set of forbidden configurations. The state of the robot is parametrized by the location h of an arbitrary point of the robot B, initially at the origin; hence the configuration space $\mathscr{C}$ is identical to $\mathbb{R}^2$. The allowed motions form the translation group $\mathbf{T}$ which can be identified with $\mathbb{R}^2$, see Example 1 above. The forbidden points can be identified with the set

$$\mathscr{D}_B(X) = \{h \in \mathbb{R}^2 : \tau_h B \quad X\}. \tag{44}$$

One immediately recognizes this (cf. (3)) as the Euclidean dilation of X by $\check{B}$. This leads to the first result.

Proposition 8. *If $B \in \mathbb{R}^2$ is a robot with translational degrees of freedom, then the hitting function $\mathscr{D}_B \colon \mathscr{P}(\mathbb{R}^2) \to \mathscr{P}(\mathscr{C})$ is given by*

$$\mathscr{D}_B(X) = X \overset{\mathbf{T}}{\oplus} \check{B}$$

$$=: \delta_B^{\mathbf{T}}(X) = \bigcup_{x \in X} \tau_x \check{B}. \tag{45}$$

For clarity the dependence of the Minkowski sum on the Euclidean translation group $\mathbf{T}$ is explicitly indicated in (45). An example can be found in Fig. 5.

4.2. Translations and Rotations

Next the case of a mobile robot with translational and rotational degrees of freedom is considered. The appropriate group is $\mathbf{M}$, the Euclidean motion group, see Example 2. To parametrize the state of the robot, choose two distinct points,

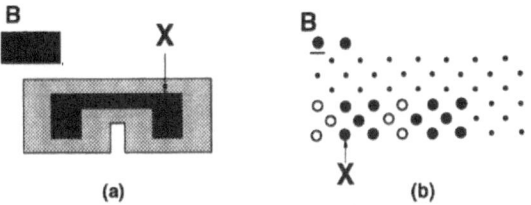

Figure 5. The forbidden set $\mathcal{D}_B(X)$ for a robot B with translational degrees of freedom. **a** Continuous case: dark area X: the obstacle; hatched area (plus the enclosed dark area): the dilated set. The origin ω is in the center of the rectangle B. **b** Discrete case: heavy dots: the obstacle X; heavy plus open dots: the dilated set. The underlining in B indicates the location of the origin

say P_1 and P_2 inside the robot. The configuration space $\mathcal{C}$ is 3-dimensional in this case,

$$\mathcal{C} := \{(h, \phi): h \in \mathbb{R}^2, \phi \in S^1\}, \tag{46}$$

with h the location of point P_1 and ϕ the angle of the line segment $[P_1 P_2]$ with respect to the x-axis. Note that $\mathcal{C}$ can be identified with the parameter space of the Euclidean motion group $\mathbf{M}$: to each (h, ϕ) in $\mathcal{C}$ corresponds a unique motion $\gamma_{h, \phi} = \tau_h \rho_\phi^0$ and vice versa. This identification will tacitly be made below without further comment. An alternative way to represent $\mathcal{C}$ is by means of pointers; see Section 2.4. Assuming the initial state to be equal to $\{h = (0, 0), \phi = 0\}$, the hitting function in this case becomes

$$\mathcal{D}_B(X) = \{(h, \phi) \in \mathcal{C}: \gamma_{h, \phi} B \quad X\}. \tag{47}$$

Now the results of Section 2.5 are applicable. From the general result (20) together with (17)–(18) it follows that

$$\mathcal{D}_B(X) = \vartheta(X) \overset{\mathbf{M}}{\oplus} \vartheta^{-1}(B)$$
$$= \vartheta(X) \overset{\mathbf{M}}{\oplus} (\mathbf{R} \overset{\mathbf{M}}{\oplus} \tau^{-1}(B)) \tag{48}$$

where (cf. (18))

$$\tau^{-1}(B) = \{\tau_b^{-1}: b \in B\} = \tau(\check{B}), \tag{49}$$

with $\check{B}$ the reflected set of B. From (48) one obtains the following result.

Proposition 9. *If $B \in \mathbb{R}^2$ is a robot with translational and rotational degrees of freedom, then the hitting function $\mathcal{D}_B: \mathcal{P}(\mathbb{R}^2) \to \mathcal{P}(\mathcal{C})$ is given by*

$$\mathcal{D}_B(X) = \bigcup_{x \in X} \tau_x(\mathbf{R} \overset{\mathbf{M}}{\oplus} \tau^{-1}(B)) \tag{50}$$

$$= \bigcup_{\phi \in S^1} \bigcup_{x \in X} \tau_x \rho_\phi^0 \tau(\check{B}) \tag{51}$$

The equality (51) expresses the fact, which is obvious from (47), that $\mathcal{D}_B(X)$ can be found by doing, for each $\phi \in S^1$ an ordinary dilation with a rotated version $\rho_\phi^0 B$ of the structuring element. Equation (50) says that $\mathcal{D}_B(X)$ can also be found as a union of translates, that is, by a dilation

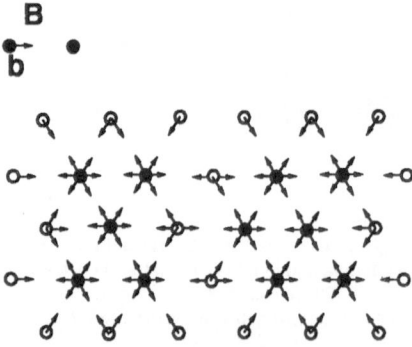

Figure 6. The forbidden set $\mathscr{D}_B(X)$ for a robot B with translational and rotational degrees of freedom. Heavy dots: the obstacle space X; b base pointer. Arrows attached to heavy and open dots: the forbidden set

$$\delta_{\check{B}}^{T}(X) := \bigcup_{x \in X} \tau_x \tilde{B}, \tag{52}$$

where

$$\tilde{B} := \mathbf{R} \overset{M}{\oplus} \tau^{-1}(B). \tag{53}$$

Eq. (52) differs from a usual T-dilation through the fact that the structuring element (53) is not a planar, but a 3D subset of $\mathscr{C}$. The construction of $\tilde{B}$ is straightforward: (i) take the reflection $\check{B}$ of B; (ii) lift $\check{B}$ to $\mathscr{C}$ by applying τ; (iii) construct rotated copies of $\tau(\check{B})$ in $\mathscr{C}$.

These results can be nicely presented geometrically using the representation by pointers, see Fig. 6. Alternatively, one may use a 3D representation in configuration space $\mathscr{C}$.

5. Conclusions

In this paper we have presented an extension of mathematical morphology with a larger symmetry group than the group of Euclidean translations. It has been shown how to obtain generalizations of morphological transformations for arbitrary homogeneous spaces.

Using this framework, one of the limitations mathematical morphology is addressed here: how to take the projective geometry of the imaging process into account. In Section 3 we explained how to define morphological operations on images of a planar patch produced by perspective projection, where the allowed 3D motion is restricted to translation within the object plane. The main result (Theorem 7) justifies existing procedures of adapting the structuring element as a function of the position in the image plane w.r.t. the vanishing line.

Finally in Section 4 a problem from robot vision for path planning was approached by morphological methods. Here the problem is to find a path for an object, say a robot or a car, moving in a space with obstacles. It was shown, both when translations and rotations of the robot are allowed, how morphological methods may be used to solve the *empty-space problem*: find the allowed states of the robot.

References

[1] Berger, M.: Geometry I. Berlin, Heidelberg, New York, Tokyo: Springer 1987.

[2] Beucher, S., Blosseville, J. M., Lenoir, F.: Traffic spatial measurements using video image processing. In: SPIE Cambridge 87 Symp. on Advances in Intelligent Robotics Systems, Nov. 1987.

[3] Gouzènes, L.: Strategies for solving collision-free trajectories problems for mobile and manipulator robots. Int. J. Robotics Res. *3*, 51–65 (1984).

[4] Hadwiger, H.: Vorlesungen über Inhalt, Oberfläche, und Isoperimetrie. Berlin, Göttingen, Heidelberg: Springer 1957.

[5] Heijmans, H. J. A. M.: Mathematical morphology: an algebraic approach. CWI Newslett. *14*, 7–27 (1987).

[6] Heijmans, H. J. A. M., Ronse, C.: The algebraic basis of mathematical morphology. Part I: dilations and erosions. Comp. Vis. Graph. Im. Proc. *50*, 245–295 (1989).

[7] Heijmans, H. J. A. M.: Morphological image operators. New York: Academic Press 1994 (Advances in Electronics and Electron Physics, Suppl. Vol. 25).

[8] Kendall, D.: Shape manifolds, procrustean metrics, and complex projective spaces. Bull. London Math. Soc. *16*, 81–121 (1984).

[9] Klein, F.: Vergleichende Betrachtungen über neuere geometrische Forschungen. In: Gesammelte mathematische Abbandlungen, Vol. I, pp. 460–497, 1872.

[10] Matheron, G.: Random sets and integral geometry. J. Wiley: New York 1975.

[11] Mundy, J. L., Zisserman, A., Forsyth, D., eds.: Applications of invariance in computer vision. Berlin, Heidelberg, New York, Tokyo: Springer 1994 (Lecture Notes in Computer Science, Vol. 825).

[12] Robinson, D. J. S.: A course in the theory of groups. Berlin, Heidelberg, New York, Tokyo: Springer 1982.

[13] Roerdink, J. B. T. M.: On the construction of translation and rotation invariant morphological operators. Report AM-R9025, Centre for Mathematics and Computer Science, Amsterdam, 1990 (To appear in: Mathematical morphology: theory and hardware, R. M. Haralick, ed., Oxford University Press).

[14] Roerdink, J. B. T. M.: Mathematical morphology on the sphere. In: Proc. SPIE Conf. Visual Communications and Image Processing '90, Lausanne, pp. 263–271, 1990.

[15] Roerdink, J. B. T. M.: Mathematical morphology with non-commutative symmetry groups. In: Mathematical morphology in image processing (Dougherty, E. R., ed.), pp. 205–254. New York: Marcel Dekker 1993.

[16] Roerdink, J. B. T. M.: Solving the empty space problem in robot path planning by mathematical morphology. In: Proc. Workshop 'Mathematical Morphology and its Applications to Signal Processing', Barcelona, Spain, May 12–14 (Serra, J., Salembier, P., eds.), pp. 216–221, 1993.

[17] Roerdink, J. B. T. M.: Manifold shape: from differential geometry to mathematical morphology. In: Shape in picture (NATO ASI Series), vol. F 126, O (Y.-L., Toet, A., Foster, D., Heijmans, H. J. A. M., Meer, P., eds.), pp. 209–223. Berlin, Heidelberg, New York, Tokyo: Springer 1994.

[18] Roerdink, J. B. T. M., Heijmans, H. J. A. M.: Mathematical morphology for structures without translation symmetry. Signal Proc. *15*, 271–277 (1988).

[19] Ronse, C.: Fourier analysis, mathematical morphology, and vision. PRLB Working Document WD54, 1989.

[20] Ronse, C., Heijmans, H. J. A. M.: The algebraic basis of mathematical morphology. Part II: openings and closings. Comp. Vis. Graph. Im. Proc. Image Understanding *54*, 74–97 (1991).

[21] Serra, J.: Image analysis and mathematical morphology. New York: Academic Press 1982.

[22] Serra, J., ed.: Image analysis and mathematical morphology, Vol. 2: theoretical advances. New York: Academic Press 1988.
[23] Suzuki, M.: Group theory. Berlin, Heidelberg, New York: Springer 1982.

J. B. T. M. Roerdink
Institute for Mathematics
and Computing Science
University of Groningen
P.O. Box 800
9700 AV Groningen
The Netherlands
e-mail: roe@cs.rug.nl

Computing Suppl. 11, 149–165 (1996)

A Variational Approach to the Design of Early Vision Algorithms

C. Schnörr, R. Sprengel, and B. Neumann, Hamburg

Abstract

A Variational Approach to the Design of Early Vision Algorithms. A mathematical model for the design of early vision processing stages is presented. The model comprises a "generic" class of abstract minimization problems from which specific nonlinear diffusion processes can be derived for various kinds of visual data. Each smoothing process results in a one-parameter family of segmentations of the underlying domain and thus provides a basis for the segmentation of "general" scenes. A wide range of numerical realizations can be implemented using standard Galerkin discretization. Theoretically, each algorithm can also be implemented as a globally convergent network using analog VLSI-hardware. The approach is illustrated by deriving nonlinear diffusion schemes for the processing of greyvalue data and for the processing of locally computed image motion data.

Key words: Image processing, computer vision, convex minimization, variational segmentation, nonlinear diffusion.

1. Overview

1.1. Introduction

The development of theoretical foundations of computational vision is still in its infancy. Unlike with traditional engineering disciplines, there are no formal tools that allow to predict the performance of a developed system with respect to a visual task in a general and changing environment. Of course, this is due to the complexity of visual tasks originating in the complexity of raw visual data. We use the term "complexity" (of visual data) to refer to the multiple physical sources that give rise to these signals as well as to the manifold *local* distributions of visual data that cannot easily be described by predefined local signal models.

Neurobiology and psychophysics provide us with numerous data concerning the early processing stages of natural visual systems. These results, however, cannot be directly used to design vision algorithms. Existing low-level computational techniques, on the other hand, do usually not reflect the complexity of visual data. It is difficult to predict, for example, in which cases a step-edge based detector will be appropriate. More general, little attention is paid to realistic models of the raw visual data itself. As a consequence, attempts to characterize the performance of approaches in terms of properties of a sufficient general class of signals are rare.

Within the last years, various new trends have emerged in the field of computer vision, concerning both technical tools to devise vision algorithms and the viewpoint of the system developer itself. On the one hand, more and more vision scientists resort to sound mathematical methods for modelling problems related with early vision. On the other hand, the active-vision viewpoint [5] has moved to the foreground problems related with the integration of processing stages and the control of systems based on task-oriented cost functionals.

In this context, this article presents a mathematical model for a specific class of nonlinear early vision algorithms. The general task we tackle is the segmentation problem, that, roughly spoken, is the problem to devise general mechanisms that endow a computer vision system with the capability to separate essential features of the visual data flow from the rest. As this capability is basic to any computer vision system, solutions to the segmentation problem are important. On the other hand, this problem is intricate in that the meaning of "essential" (features) varies with the context and is difficult to formalize along with an appropriate description of general visual data.

Our contribution reflects the remarks made above as follows:

1) As a "model" for image data we only assume these data to be samples of a square-integrable function. In contrast to many approaches using explicit feature models, we do not assume any local properties *a priori* but merely take into account perturbations of the signal by sensor noise and quantization effects.

2) As a model for the processing stage we construct a nonlinear mapping from the input data space to another space of functions. This mapping is the inverse of a potential operator that is derived from an optimality principle favouring piecewise smooth functions as solutions. Existence and uniqueness of these solutions are shown. Thus our contribution concerns a sound mathematical model for non-local cooperative processes that, by virtue of its nonlinearity, makes the various local structure of visual input data visible for subsequent processing stages.

3) Galerkin discretization converts our model to a system of nonlinear ordinary differential equations describing the activation dynamics of a globally asymptotically stable network. Thus, in principle, a physical implementation with analogue hardware is feasible and would result in a stable real time processing stage. Alternatively, standard numerical methods lead to massively parallel algorithms for the computation of approximate solutions. Convergence of these approximations to the continuous solutions holds true. Hence, irrespective of discretization details numerically computed solutions inherit all properties (and only those) of the underlying continuous formulation.

4) Our model involves two global parameters: One of them accounts for the unknown scale of the relevant structures of the input data. The other parameter controls the sensivity of the process to image structures or, with other words, the amount of information made available to subsequent stages. We regard the question for "the best" values of these parameters as irrelevant. Rather we

regard these parameters as active components of our model, the control of which has to be considered along with models for a complete segmentation stage, dependend on the high-level state of the system to which it belongs. As a necessary prerequisite we show that the results of our processing stage smoothly depend on variations of the model parameters.

5) Our model does not depend on the kind (greyvalues, color, motion, ...) of the input data. Consequently, the same principle can be used to design processes for diverse visual properties. Examples for greyvalue data and motion data are given later on. Supported by biological findings, the use of such processes in parallel along with a corresponding integration stage appears to be necessary to compute stable segmentations from general visual data.

1.2. Related Work

Variational approaches to the segmentation problem has been a topic of intensive research during the last decade. For a review we refer to [35].

In contrast to our continuously formulated approach, almost all approaches in the computer vision literature are discrete. They are based on models that use two sets of variables arranged on uniform meshes: One set representing a piecewise smooth version of the image data and a second set of binary variables indicating discontinuities (e.g. [7–9, 17, 18, 25, 27]). The resulting global minimization problems are very complex, and since appropriate simulated annealing procedures are extremely slow in practice most researchers compute a discrete sequence of local minima by minimizing a related one-parameter family of cost-functionals (see [8, 17]). As a result, this procedure comprises many parameters, and it is not clear (i) under which conditions a "good" local minimum can be reached and (ii) whether the discontinuity set continuously depends on the given image data.

An advantage of continuously formulated approaches to the segmentation problem is that they do not depend on the image coordinate system used, and that the problem to compute shapes from image data can be treated in a non-discrete way. Mumford and Shah [26] presented a variational approach to the segmentation problem that triggered research in mathematical fields as well [3, 15]. However, as illustrated by recent work in computer vision [22, 28], there appears no theory to be available that allows to approximate and to realize the approach of Mumford and Shah numerically. For a special case of the Mumford and Shah functional, however, a numerical approach has been recently presented [21].

A third line of research concerns nonlinear diffusion approaches [1, 2, 12, 29, 31]. These approaches are closer to our approach and will be discussed in Section 5.2.

1.3. Outline of the Paper

The mathematical basis of our approach is described in Section 2.1 and applied to the processing of grevvalue images and visual motion in Sections 2.2 and 4.1, respectively. Section 2.3 indicates how a wide range of numerical realizations can be obtained by standard methods, and properties of our approach are illustrated by numerical examples in Sections 3 and 4.2. We discuss our results in Section 5.1 and related approaches in Section 5.2. Further research is indicated in Section 5.3.

1.4. Notation

For Hilbert spaces and related concepts of functional analysis we refer to, e.g., [4]. $L^2(\Omega; \mathbb{R}^n)$ denotes the Hilbert space of measurable and square-integrable functions $u: \Omega \to \mathbb{R}^n$ with scalar product

$$(u, v)_0 = \int_\Omega u \cdot v dx,$$

and norm

$$\|u\|_0 = (u, u)_0^{1/2}.$$

We write $L^2(\Omega)$ for $L^2(\Omega; \mathbb{R}^1)$.

$H^1(\Omega)$ denotes the Sobolev space

$$H^1(\Omega) = \left\{ u \in L^2(\Omega) \,\middle|\, \frac{\partial}{\partial x_i} u \in L^2(\Omega), i = 1, \ldots, n \right\}$$

with scalar product

$$(u, v)_1 = (u, v)_0 + (Du, Dv)_0$$

and norm

$$\|u\|_1 = (u, u)_1^{1/2}.$$

Here (and below), the symbol D is used for the gradient operator with respect to the spatial variables,

$$Du = \left(\frac{\partial}{\partial x_1} u, \ldots, \frac{\partial}{\partial x_n} u \right)^t,$$

and the derivatives are to be understood in the sense of distributions.

The dual space of a Hilbert space $\mathcal{H}$ is denoted with $\mathcal{H}^*$. For the value $f(u)$ of a linear functional $f \in \mathcal{H}^*$ applied to some element $u \in \mathcal{H}$ we write

$$\langle f, u \rangle.$$

2. Approach

2.1. Mathematical Model

Our approach is based on the following mathematical model: The image data are considered to represent an element of a Hilbert space $\mathcal{H}_1$, and we seek a piecewise smooth version of these data as an unique element u of a Hilbert space $\mathcal{H}_2$ by minimizing a C^1-functional

$$J: \mathcal{H}_2 \to \mathbb{R}, \qquad J: v \to J(v) = F(v) - f(v) + const. \tag{1}$$

The functional J comprises a non-quadratic term $F(\cdot)$, a linear functional $f(\cdot)$ involving the given image data and possibly a constant (see Sections 2.2 and 4.1 for concrete examples). Taking the derivative of J, u satisfies the operator equation

$$A(u) = f, \tag{2}$$

where $A = F'$. Existence and uniqueness of u is achieved by modelling the functional $F(\cdot)$ in (1) such that the operator $A: \mathcal{H}_2 \to \mathcal{H}_2^*$ is strongly monotone (see [10, 38]), that is

$$\langle A(v) - A(w), v - w \rangle \geq c_m \|v - w\|_{\mathcal{H}_2}^2, \qquad c_m > 0, \qquad \forall v, w \in \mathcal{H}_2, \tag{3}$$

and Lipschitz-continuous:

$$\|A(v) - A(w)\|_{\mathcal{H}_2^*} \leq c_L \|v - w\|_{\mathcal{H}_2}, \qquad c_L > 0, \qquad \forall v, w \in \mathcal{H}_2. \tag{4}$$

Moreover, approximate solutions $u_h \in \mathcal{H}_2$ belonging to an appropriate Finite-Element subspace converge to u as the discretization parameter h goes to zero:

$$\|u - u_h\|_{\mathcal{H}_2} \to 0 \qquad for \qquad h \to 0.$$

2.2. Determining Piecewise Smooth Functions

We apply the framework outlined above to the prototype problem to determine a piecewise smooth version of given image data $d: \Omega \subset \mathbb{R}^n \to \mathbb{R}$. To this end, we assume $d \in L^2(\Omega)$ and consider the functional

$$J: H^1(\Omega) \to \mathbb{R}, \qquad J: v \to J(v) = \int_\Omega \{(v - d)^2 + \lambda(Dv)\} \, dx, \tag{5}$$

where $\lambda(\cdot): \mathbb{R}^n \to \mathbb{R}$ is given by

$$\lambda(x) = \psi(|x|)$$

$$\psi(t) = \begin{cases} \lambda_h^2 t^2, & 0 \leq t \leq c_\rho \\ \lambda_l^2 t^2 + 2c_\psi t - c_\psi c_\rho, & 0 \leq c_\rho \leq t \end{cases}$$

$$c_\psi = (\lambda_h^2 - \lambda_l^2) c_\rho; \qquad \lambda_h, \lambda_l, c_\rho > 0. \tag{6}$$

The functional J in (5) comprises two terms measuring the deviation of functions v from the data d and the smoothness of v. The smoothness term involves the function $\lambda(\cdot)$ which, by definition (6), quadratically depends on the magnitude of the gradient $|Dv|$ for $|Dv| \leq c_\rho$, and almost linearly for $|Dv| \geq c_\rho$ (λ_l, which we always set to a small value, is only of theoretical relevance). As a result, any algorithm minimizing J adaptively smooths the data d and tends to preserve significant transitions of d. This approach can be considered as a natural extension of a class of linear non-local mechanisms that has been considered as a paradigma of early computational vision [32], to the nonlinear case (see [36]).

The variational formulation of Eq. (2),

$$\langle A(u), v \rangle = \langle f, v \rangle, \qquad \forall v \in H^1(\Omega), \tag{7}$$

reads

$$\int_\Omega \{2uv + \lambda'(Du) \cdot Dv\} \, dx = 2 \int_\Omega dv \, dx, \qquad \forall v \in H^1(\Omega), \tag{8}$$

and it remains to verify conditions (3) and (4). By virtue of the inequalities (see Definition 6)

$$(\lambda'(x) - \lambda'(y)) \cdot (x - y) \geq 2\lambda_i^2 |x - y|^2, \qquad \forall x, y \in \mathbb{R}^n,$$

$$|\lambda'(x) - \lambda'(y)| \leq 2\lambda_h^2 |x - y|, \qquad \forall x, y \in \mathbb{R}^n, \tag{9}$$

we obtain from (7) and (8)

$$\langle A(u) - A(v), u - v \rangle \geq c_m \int_\Omega \{(u - v)^2 + |Du - Dv|^2\} \, dx$$

$$= c_m \|u - v\|_1^2, \qquad c_m = \min\{2, 2\lambda_i^2\},$$

and

$$|\langle A(u) - A(v), w \rangle| \leq 2\|u - v\|_0 \|w\|_0 + 2\lambda_h^2 \|Du - Dv\|_0 \|Dw\|_0$$

$$\leq c_L \|u - v\|_1 \|w\|_1, \qquad c_L = \max\{2, 2\lambda_h^2\}.$$

2.3. Algorithms

Approximate solutions u_h are computed by restricting Eq. (8) to a finite-dimensional subspace $\mathscr{H}_h \subset H^1(\Omega)$ using piecewise-linear Finite Elements [13]. Let $\{\phi_i \in \mathscr{H}_h\}$, $i = 1 \ldots n$, denote the set of basis functions according to a triangulation of the underlying domain Ω, and let P be the isomorphism

$$P: \mathbb{R}^n \to \mathscr{H}_h, \qquad P: \mathbf{v} \to P\mathbf{v} = u_h = \sum_{i=1}^n v_i \phi_i.$$

Substitution in equation (8) yields a set of nonlinear equations

$$(\mathbf{G}(\mathbf{u}) - \mathbf{g}) \cdot \mathbf{v} = \langle A(P\mathbf{u}) - f, P\mathbf{v} \rangle = 0, \forall \mathbf{v} \in \mathbb{R}^n,$$

that has a unique root $\mathbf{u}$. The nonlinear mapping $\mathbf{G}: \mathbb{R}^n \to \mathbb{R}^n$ inherits the monotonicity (3) and Lipschitz continuity (4) of the operator A, and based on the inequality

$$|\mathbf{G}(\mathbf{v}) - \mathbf{G}(\mathbf{w})| \leq \tilde{c}_L |\mathbf{v} - \mathbf{w}|, \qquad \tilde{c}_L = \tilde{c}_L(c_L, \|P\|), \qquad \forall \mathbf{v}, \mathbf{w} \in \mathbb{R}^n,$$

u_h can be computed by an appropriate iterative algorithm [30]. To date, we have implemented an iterative Jacobi-solver on a vector-computer. More efficient multigrid-solvers will be developed in future work.

An important question for the purposes of early vision is whether—in principle—a realization of an algorithm using analog VLSI is feasible. To this end, we regard the set of nonlinear differential equations

$$\frac{d}{dt}\mathbf{v}(t) = -(\mathbf{G}(\mathbf{v}(t)) - \mathbf{g}) \tag{10}$$

as description of the activation dynamics $\mathbf{v}(t)$ of a "neural network" that comprises the components $\mathbf{v}_i$ of $\mathbf{v}$ as computational units and receives clamped stimuli $\mathbf{g}_i$ according to the data $d(x)$. By virtue of condition (3), it can be shown [36] that, for any trajectories $\mathbf{v}(t)$, $\mathbf{w}(t)$ in state space,

$$\frac{d}{dt}|\mathbf{v}(t) - \mathbf{w}(t)|^2 \le c|\mathbf{v} - \mathbf{w}|^2, \qquad c = c(c_m, \|P^{-1}\|).$$

Thus, trajectories uniformly tend to the unique global equilibrium $\mathbf{u}$ and, according to a classification of Hirsch [19], the network described by (10) is globally asymptotically stable.

3. Properties and Numerical Examples

In this Section we first exhibit in more detail how the nonlinear smoothing mechanism works locally. Next we present numerical examples illustrating the main features of our approach. Finally, a comparison is carried out with classical edge-detection approaches known from the literature.

3.1. Smoothing Behaviour

A better understanding of the smoothing behaviour of the approach is gained by considering the Euler-Lagrange equations attached to the minimization problem (5). Introducing the diffusion-coefficient

$$\rho(t) = \frac{\psi'(t)}{t}, \qquad t \ge 0, \tag{11}$$

partial integration in (8) shows that minimizing solutions may be considered as (generalized) steady-state solutions to the nonlinear diffusion equation

$$\frac{dv}{dt} = D \cdot (\rho(|Dv|) Dv) - 2(v - d), \qquad \frac{dv}{dn} = 0 \quad \text{on} \quad \partial\Omega. \tag{12}$$

Expanding the smoothing term of this equation and rearranging, we obtain (H denotes the Hessian of v):

$$\begin{aligned}
D \cdot (\rho|Dv|) Dv) &= D\rho(|Dv|) \cdot Dv + \rho(|Dv|)\Delta v \\
&= \frac{\rho'(|Dv|)}{|Dv|} Dv \cdot HDv + \rho(|Dv|)\Delta v \\
&= (\rho'(|Dv|)|Dv| + \rho(|Dv|))\left(\frac{d^2}{de_1^2}v\right) + \rho(|Dv|)\left(\frac{d^2}{de_2^2}v\right)
\end{aligned} \tag{13}$$

where

$$e_1 = \frac{Dv}{|Dv|}, \quad \text{and} \quad e_2 = \frac{(Dv)^\perp}{|Dv|}.$$

According to the definition of $\rho(\cdot)$ (11) and $\psi(\cdot)$ (6), the two terms in Eq. (13) specifying the amount of smoothing in the directions e_1, e_2 explicitly read:

$$\rho'(t)t + \rho(t) = \begin{cases} 2\lambda_h^2, & t < c_\rho \\ 2\lambda_l^2, & t > c_\rho \end{cases},$$

and

$$\rho(t) = \begin{cases} 2\lambda_h^2, & t < c_\rho \\ 2\lambda_l^2 + 2c_\psi/t, & t > c_\rho \end{cases}.$$

Thus, Eq. (13) says that along the gradient direction Dv the diffusion process abruptly slows down (recall that $0 < \lambda_l \ll 1$), whereas smoothing along the level-curves of v only decreases $\sim |Dv|^{-1}$.

This result shows that the global optimization problem (5) leads to a cooperative nonlinear process that performs adaptive smoothing of local image structure in a very plausible way.

3.2. Numerical Examples

To demonstrate the generality of the approach, we show two examples from two quite different domains: Medical imaging and scenes with man-made objects.

In Section 2.2 we explained that algorithms minimizing the functional J in (5) adaptively preserve transitions of the input data, and that the sensivity with respect to transitions is controlled by the parameter c_ρ. Accordingly, after convergence of the algorithm to the unique minimizing function u, we obtain a partition of the domain Ω:

$$\Omega = \Omega_l \cup \Omega_h, \qquad \Omega_l = \{x \in \Omega : |Du(x)| > c_\rho\}. \tag{14}$$

Figure 1 shows two real images of different application areas. Figures 2 and 3 show the sets Ω_l for decreasing values of the parameter c_ρ (from left to right).

The following observations can be made:

- The amount of structure of the input data that is detected is controlled by the global parameter c_ρ. Lower values of this parameter make weaker transitions visible. Apparently we have $\Omega_l \subset \Omega_l'$ for $c_\rho > c_\rho'$.
- The transition regions Ω_l exhibit interesting topological properties. In particular, important clues for a 3D-interpretation or registration techniques like triple-points, that is junctions where three edges meet each other, are preserved. This is in contrast to traditional local transition detectors like the Laplacian of Gaussian, for example (see next Section).

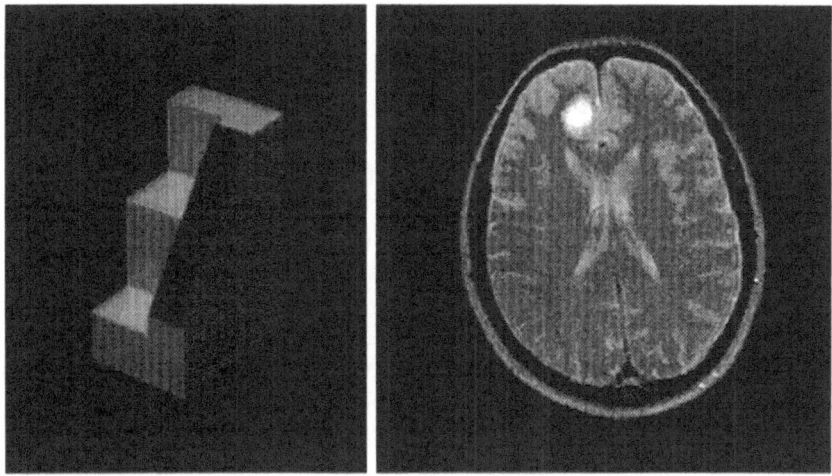

Figure 1. Image of a man-made object and computer tomogram

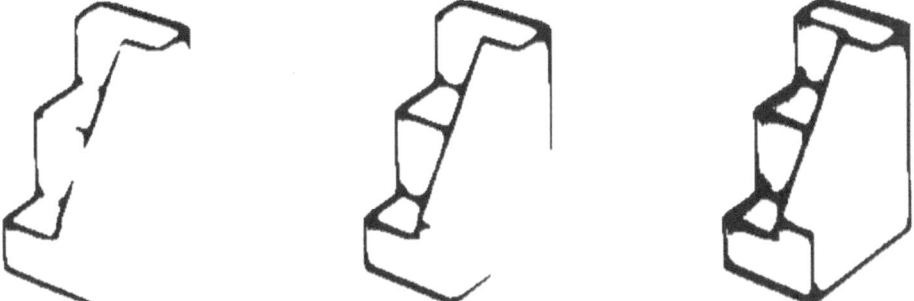

Figure 2. Computed transition-regions of object-image. *Left* $\lambda_h = 7$, $c_\rho = 4$, *middle* $\lambda_h = 7$, $c_\rho = 3$, *right* $\lambda_h = 7$, $c_\rho = 1.5$

Figure 3. Computed transition-regions of computer tomogram. *Left* $\lambda_h = 9$, $c_\rho = 0.8$, *middle* $\lambda_h = 9$, $c_\rho = 0.6$, *right* $\lambda_h = 9$, $c_\rho = 0.36$

- The localization of transitions is stable with respect to the variation of c_ρ. Again, this is in contrast to traditional linear techniques where representations of data transitions move in "scale-space".

3.3. Comparison with Traditional Techniques

In this Section, we contrast our results (i) with zero-crossings of Laplacian (of Gaussian)-filtered images [24, 23], and (ii) Canny's edge-detection method [11].

Figure 4 shows the zero-crossings of the tomogram-image in Fig. 1 that was filtered with the Laplacian-of-Gaussian at three different scales σ. The main feature of this approach is that the curves obtained are always closed. In contrast to the sets depicted in Fig. 3 one observes, however, (i) that the zero-crossings move in "scale-space", and (ii) that—concerning localization and topology—these curves do not very well represent most of the anatomic structures.

Figure 5 (left) shows the locations of the Gaussian-filtered tomogram-image where the gradient exceeds some threshold. This method may be considered, to some

Figure 4. Zero-crossings of the Laplacian-of-Gaussian. *Left $\sigma = 1.0$, middle $\sigma = 2.0$, right $\sigma = 4.0$*

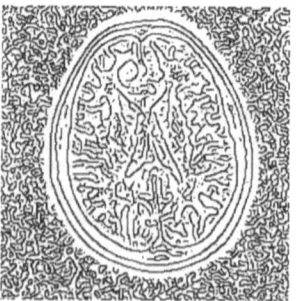

Figure 5. Canny's edge-detection approach (see text). *Left $\sigma = 1.5$, $|D(G_\sigma * u)| > 3.5$, middle $\sigma = 1.5$, nms, right $\sigma = 1.5$, nms, hi $= 3.0$*

extent, as a linear version of our approach. In order to have comparable results, the threshold was manually adjusted such that the subset corresponding to the boundary of the ventricel system (in the middle of the image) is nearly closed. A comparison with the middle image in Fig. 3 reveals (i) that the structures shown in Fig. 5 (left) are much broader and tend to merge, and (ii) that more spurious details are detected.

Finally, Fig. 5 (middle and right) shows the left image further processed with non-maximum suppression (nms) and hysteresis thresholding. However, the edges obtained in this way do not preserve important topological properties like the contour of the ventrical system and the tumor, for example.

4. Application to the Processing of Visual Motion

In Section 1.1, we have claimed the applicability of our framework to various problem areas of early vision. Here, we outline a first attempt to apply our approach to the processing of visual motion.

4.1. Approach

Visual clues induced by movements of the sensor relative to the environment is a major source of information for vision-based systems. A common technique for the first processing stage is to extract "conserved quantities" $\phi(x, t)$ by locally filtering raw image data [20, 37, 39] and to evaluate constraint equations of the form

$$\frac{d}{dt}\phi = D\phi \cdot v + \phi_t = 0, \tag{15}$$

where

$$v = \begin{pmatrix} v1 \\ v2 \end{pmatrix} = \frac{d}{dt}x = \begin{pmatrix} \dot{x}1 \\ \dot{x}2 \end{pmatrix}$$

is the so-called motion field, that is the vector field of instantaneous velocities of projected scene points within the image plane. Sharp transitions of motion fields usually arise at image regions that correspond to boundaries of moving objects, for example. Therefore, the ability to detect such regions while smoothing out noise is important. To this end, we wish to formulate a suitable minimization problem along the lines of Section 2.1.

Let

$$Lv = b \tag{16}$$

be a linear system of equations of the form (15). An approach according to the general model of Section 2.1 and analoguous to the prototype problem described in Section 2.2 is:

$$J: \mathcal{H} := H^1(\Omega) \times H^1(\Omega) \to \mathbb{R},$$

$$J(v) = \int_\Omega \left\{ |L_\alpha^+(Lv - b)|^2 + \frac{1}{2}(\lambda_d(div(v)) + \lambda_r(rot(v)) + \lambda_s(sh(v))) \right\} dx, \quad (17)$$

where

$$L_\alpha^+ = (L^t L + \alpha I)^{-1} L^t, \qquad \alpha > 0. \tag{18}$$

$\lambda_d(\cdot)$, $\lambda_r(\cdot)$ and $\lambda_s(\cdot)$ are defined as $\lambda(\cdot)$ in (6), and

$$div(v) = v1_{x1} + v2_{x2}, \quad rot(v) = v2_{x1} - v1_{x2}, \quad sh(v) = \begin{pmatrix} v2_{x2} - v1_{x1} \\ v1_{x2} + v2_{x1} \end{pmatrix}. \tag{19}$$

Disregarding the regularizing constant α, the first term of the functional J in (17) compares the least-squares solution $L_\alpha^+ b$ to the system (16) with the corresponding component of v in the orthogonal complement of the null-space of L. The other terms embody nonlinear smoothing processes for each component of the vector-gradient of v.

Equation (2) corresponding to the minimization problem (17) now reads

$$\langle A(u), v \rangle = \int_\Omega \left\{ 2(L_\alpha^+ Lu) \cdot (L_\alpha^+ Lv) \right.$$

$$\left. + \frac{1}{2}(\lambda_d'(div(u)) \, div(v) + \lambda_r'(rot(u)) \, rot(v) + \lambda_s'(sh(u)) \cdot sh(v)) \right\} dx$$

$$= 2 \int_\Omega (L_\alpha^+ Lv) \cdot L_\alpha^+ b \, dx, \qquad \forall v \in \mathcal{H}. \tag{20}$$

To verify conditions (3) and (4), we introduce the notation

$$(v, w)_{\mathcal{H}} = (v1, w1)_1 + (v2, w2)_1$$

$$\|v\|_{\mathcal{H}} = (v, v)_{\mathcal{H}}^{1/2}$$

$$|g|_\infty = \|g\|_{L^\infty(\Omega)}, \qquad \forall g \in L^\infty(\Omega)$$

$$\begin{pmatrix} a & b \\ b & c \end{pmatrix} := L_\alpha^+ L$$

and assume $a(x), b(x), c(x) \in L^\infty(\Omega)$. By virtue of (9), we obtain from (20)

$$\langle A(u) - A(v), u - v \rangle \geq \int_\Omega \left\{ 2[a(u1 - v1) + b(u2 - v2)]^2 \right.$$

$$+ 2[b(u1 - v1) + c(u2 - v2)]^2 + \lambda_{d,l}^2(div(u) - div(v))^2$$

$$\left. + \lambda_{r,l}^2(rot(u) - rot(v))^2 + \lambda_{s,l}^2 |sh(u) - sh(v)|^2 \right\} dx$$

$$\geq \int_\Omega \left\{ 2[a(u1 - v1) + b(u2 - v2)]^2 \right.$$

$$+ 2[b(u1 - v1) + c(u2 - v2)]^2 + 2 \min\{\lambda_{d,l}^2, \lambda_{r,l}^2, \lambda_{s,l}^2\}$$

$$\left. \times (|D(u1 - v1)|^2 + |D(u2 - v2)|^2) \right\} dx.$$

Putting $w = u - v$, it remains to show that

$$2 \int_{\Omega} \{(aw1 + bw2)^2 + (bw1 + cw2)^2 + \min\{\lambda_{d,l}^2, \lambda_{r,l}^2, \lambda_{s,l}^2\}(|Dw1|^2 + |Dw2|^2)\} \, dx$$

$$\geq c \|w\|_{\mathscr{H}}^2, \qquad \forall w \in \mathscr{H}.$$

This has been shown in ([34], p. 1078) based on the weak assumption that $a(x)$, $b(x)$ and $b(x)$, $c(x)$ are linear independent as elements of $L^2(\Omega)$.

To verify condition (4), we obtain from (20) using (9)

$$|\langle A(u) - A(v), w \rangle| \leq 2 \|L_{\alpha}^+ L(u - v)\|_0 \|L_{\alpha}^+ Lw\|_0$$

$$+ \frac{1}{2}(\|\lambda_d'(div(u)) - \lambda_d'(div(v))\|_0 \|div(w)\|_0$$

$$+ \|\lambda_r'(rot(u)) - \lambda_r'(rot(v))\|_0 \|rot(w)\|_0$$

$$+ \|\lambda_s'(sh(u)) - \lambda_s'(sh(v))\|_0 \|sh(w)\|_0)$$

$$\leq 8 \max\{|a^2|_{\infty}, |b^2|_{\infty}, |c^2|_{\infty}\} \|u - v\|_0 \|w\|_0$$

$$+ \max\{\lambda_{d,h}^2, \lambda_{r,h}^2, \lambda_{s,h}^2\}(\|div(u - v)\|_0 \|div(w)\|_0$$

$$+ \|rot(u - v)\|_0 \|rot(w)\|_0 + \|sh(u - v)\|_0 \|sh(w)\|_0)$$

$$\leq c \|u - v\|_{\mathscr{H}} \|w\|_{\mathscr{H}},$$

$$c = 2 \max\{8|a^2|_{\infty}, 8|b^2|_{\infty}, 8|c^2|_{\infty}, \lambda_{d,h}^2, \lambda_{b,h}^2, \lambda_{s,h}^2\}.$$

Note that, though we are now concerned with determining vector fields from image sequences, we have modeled a minimization problem that mathematically is equivalent to the prototype problem in Section 2.2.

4.2. Numerical Example

We demonstrate by a first numerical example that the observations made in Section 3 also hold true for the approach (17). The potential of this approach for visual motion processing is currently under investigation.

Figure 6 (left), shows an image of a sequence with a moving object. According to the discussion of Eq. (15), we computed a "conserved" quantity by convolving the image sequence with a set of Gaussian filters that were shifted in 3D-frequency space as described in [16]. For each node of the mesh used to discretize Eq. (20), a particular filter is selected according to a local criterion [6], and the phase $\phi(x, t)$ of the filter response is used to compute the velocity component normal to the level curves of ϕ by Eq. (15). Accordingly, the first term of the functional J in (17) explicitly reads

$$\int_{\Omega} \left(\frac{|D\phi|}{|D\phi|^2 + \alpha}(D\phi \cdot v + \phi_t) \right)^2 dx.$$

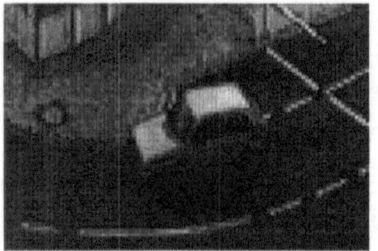 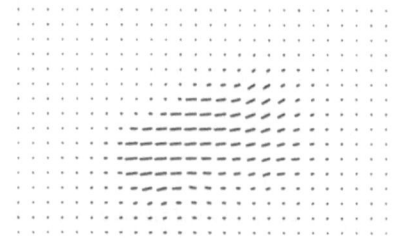

Figure 6. *Left* Image of a moving object. *Right* Estimated motion vector field

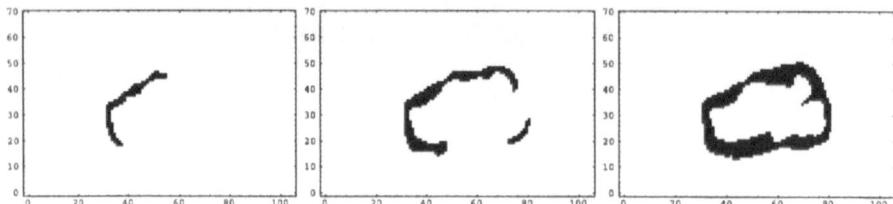

Figure 7. Computation of "motion boundaries" for the image shown in Fig. 6 (see text)

Figure 7 shows from left to right the motion-based segmentation of the image of Fig. 6 with the approach (17) for increasing sensivity of the smoothing terms in (17). Points in the image plane have been marked where the vector gradient of the resulting velocity field u,

$$(|Du1|^2 + |Du2|^2)^{1/2} = (\tfrac{1}{2}(div^2(u) + rot^2(u) + |sh(u)|^2))^{1/2},$$

exceeded the threshold $(3/2)^{1/2}c_\rho$. Again, this detection criterion directly depends on the control parameter c_ρ of the nonlinear smoothing process (cf. (14)). The (subsampled) vector field from which the last segmentation has been computed is shown in Fig. 6 (right).

Note that the nonlinear smoothing process achieves both the computation of a motion vector field by interpolating locally estimated normal velocities according to (15), and a decision about where motion boundaries occur. The results shown in Figure 7 show that the properties stated in Section 3 apply again and thus reflect the common mathematical model of the approaches (5) and (17).

5. Discussion, Related Approaches and Further Work

5.1. Discussion

Let us briefly summarize our results: We presented an approach to the segmentation problem of early vision, based on a class of nonlinear mappings between spaces of functions (Section 2.1) that guides the design of useful minimization problems for various tasks (Sections 2.2 and 4.1). For each problem, there exist numerical approximation schemes for which consistency and stability are guaranteed. Furthermore, real-time realizations using analog VLSI-hardware are in principle feasible.

Apart from the mathematical model, a characteristic feature of our approach is the design of the function $\lambda(\cdot)$ in (6) that defines two "working modes" of corresponding smoothing algorithms and, in turn, a family of segmentations of the underlying domain according to (14). The results shown in Figs. 2, 3 and 7 do obviously not correspond to "line drawings" of the underlying images that usually are the goals of edge-detection approaches, but for which yet no well-defined algorithm is known. Rather topologically correct regions are computed solely caused by transitions of input data. No attempt is made to enhance these transitions, and no implicit local signal model is used that may lead to spurious results. The complexity of the result shown in Fig. 3, for example, thus simply reflects the complexity of the corresponding image data.

5.2. Related Approaches

Nordström [29] presented an approach quite close to ours. He proposed a minimization problem similar to (5) that may lead to an enhancement of edges (diffusion locally running backward in time). However, no mathematical model seems to be available for which Nordström's approach could be shown to be well-posed. Also, one may question whether the ability to enhance detected image structures is necessary for recognition tasks.

Another related line of research concerns the design of nonlinear smoothing schemes that are not derived from an optimality principle [1, 2, 12, 31]. These approaches correspond to parabolic equations like Eq. (12) with the term $(v - d)$ omitted, $v(t = 0) = d(x)$, and right-hand sides that in general are not divergence expressions. The first approach of Perona and Malik [31] suffers like Nordström's approach from the drawbacks mentioned above, and mathematically more valid versions of that approach have been presented in [2, 12]. An important novel viewpoint has then been introduced by Alvarez et al. [1]: The design of smoothing terms that are invariant with respect to both rescalings of the image data and various transformations of the underlying domain up to projectivities of the projective plane (note that such transformations arise from looking at a plane with a pinhole-camera from different viewpoints, for example). Apart from its geometrical interpretation, the strength of this work is due to the existence of generalized solutions to the underlying evolution equations in the so-called "viscosity sense" [14]. Drawbacks of this class of approaches may arise from the fact that, in contrast to our approach, the solution functions become more and more "featureless" as $t \to \infty$. In the affine-invariant case, for example, the level curves of initial data are asymptotically deformed to ellipses and then shrink to a point [33]. Also, these approaches tacitly assume that level sets of image data comprise relevant shape information which, in general, is certainly not the case.

5.3. Further Work

Two research directions will be pursued: Firstly, we wish to understand the families of shapes that can be generated with our approach, like those of Figs. 2 and 3. This concerns questions like: What shape information can be inferred from local smoothing behaviour? Or, related to that: How can task-oriented mechanisms be devised to control smoothing behaviour? Secondly, we will apply our framework to various fields of computational vision and study problems that are specific to these areas. In the case of visual motion, for example, we expect that the adaptive behaviour of the smoothing terms in (17) will be of advantage in case of scenes that are more complex than that depicted in Fig. 6.

References

[1] Alvarez, L., Guichard, F., Lions, P. L., Morel, J. M.: Axioms and fundamental equations of image processing. Technical Report 9231, CEREMADE, Universite Paris IX—Dauphine, March 1992.
[2] Alvarez, L., Lions, P. L., Morel, J. M.: Image selective smoothing and edge detection by nonlinear diffusion. ii. SIAM J. Numer. Anal. 29, 845–866 (1992).
[3] Ambrosio, L., Tortorelli, V. M.: Approximation of functionals depending on jumps by elliptic functionals via γ-convergence. Comm. Pure Appl. Math. 43, 999–1036 (1990).
[4] Aubin, J. P.: Applied functional analysis. New York: J. Wiley 1979.
[5] Bajcsy, R.: Active perception. Proc. IEEE 76, 996–1005 (1988).
[6] Barron, J. L., Fleet, D. J., Beauchemin, S. S.: Performance of optical flow techniques. Int. J. Comput. Vision 12, 43–77 (1994).
[7] Black, M. J.: Recursive non-linear estimation of discontinuous flow fields. In: Computer Vision—ECCV '94 (Eklundh, J. O., ed.), pp. 138–145. Berlin, Heidelberg, New York, Tokyo: Springer 1994 (Lecture Notes in Computer Science, Vol. 800).
[8] Blake, A., Zisserman, A.: Visual reconstruction. Cambridge/Mass.: MIT Press 1987.
[9] Bouthemy, P., Francois, E.: Motion segmentation and qualitative dynamic scene analysis from an image sequence. Int. J. Comp. Vision 10, 157–182 (1993).
[10] Browder, F. E.: Existence theorems for nonlinear partial differential equations, part i. In: Global analysis, volume XVI of Proc. Symp. Pure Math., pp. 1–60, Providence, Rhode Island, 1970.
[11] Canny, J.: A computational approach to edge detection. IEEE Trans. Patt. Anal. Mach. Intell. 8, 679–698 (1986).
[12] Catte, F., Lions, P.-L., Morel, J.-M., Coll, T.: Image selective smoothing and edge detection by nonlinear diffusion. SIAM J. Numer. Anal. 29, 182–193 (1992).
[13] Ciarlet, P. G.: The finite element method for elliptic problems. Amsterdam: North-Holland 1978.
[14] Crandall, M. G., Ishii, H., Lions, P. L.: User's guide to viscosity solutions of second order partial linear differential equations. Bull. Amer. Math. Soc. 27, 1–67 (1992).
[15] De Giorgi, E., Carriero, M., Leaci, A.: Existence theorems for a minimum problem with free discontinuity set. Arch. Rat. Mech. Anal. 108, 195–218 (1989).
[16] Fleet, D. J., Jepson, A. D.: Computation of component image velocity from local phase information. Int. J. Comp. Vision 5, 77–104 (1990).
[17] Geiger, D., Yuille, A.: A common framework for image segmentation. Int. J. Comp. Vision 6, 227–243 (1991).
[18] Geman, S., Geman, D.: Stochastic relaxation, gibbs distributions, and the bayesian restoration of images. IEEE Trans. Patt. Anal. Mach. Intell. 6, 721–741 (1984).
[19] Hirsch, M. W.: Convergent activation dynamics in continuous time networks. Neural Networks 2, 331–349 (1989).
[20] Horn, B. K. P., Schunck, B. G.: Determining optical flow. Art. Intell. 17, 185–203 (1981).
[21] Koepfler, G., Lopez, C., Morel, J.: A multiscale algorithm for image segmentation by variational method. SIAM J. Numer. Anal. 31, 282–299 (1994).
[22] March, R.: Visual reconstruction with discontinuities using variational methods. Image Vis. Comp. 10, 30–38 (1992).
[23] Marr, D.: Vision. San Francisco: W. H. Freeman 1982.

[24] Marr, D., Hildreth, E.: Theory of edge detection. Proc. R. Soc. London Ser. *B207*, 187–217 (1980).
[25] Marroquin, J., Mitter, S., Poggio, T.: Probabilistic solution of ill-posed problems in computational vision. J. Amer. Stat. Assoc. *82*, 76–89 (1987).
[26] Mumford, D., Shah, J.: Optimal approximations by piecewise smooth functions and associated variational problems. Comm. Pure Appl. Math. *42*, 577–685 (1989).
[27] Murray, D. W., Buxton, B. F.: Scene segmentation from visual motion using global optimization. IEEE Trans. Patt. Anal. Mach. Intell. *9*, 220–228 (1987).
[28] Nesi, P.: Variational approach to optical flow estimation managing discontinuities. Image Vis. Comp. *11*, 419–439 (1993).
[29] Nordström, N.: Biased anisotropic diffusion—a unified regularization and diffusion approach to edge detection. Image Vis. Comp. *8*, 318–327 (1990).
[30] Ortega, J. M., Rheinboldt, W. C.: Iterative solution of nonlinear equations in several variables. New York: Academic Press 1970.
[31] Perona, P., Malik, J.: Scale-space and edge-detection. IEEE Trans. Patt. Anal. Mach. Intell. *12*, 629–639 (1990).
[32] Poggio, T., Torre, V., Koch, C.: Computational vision and regularization theory. Nature *317*, 314–319 (1985).
[33] Sapiro, G., Tannenbaum, A.: Affine invariant scale-space. Int. J. Comput. Vision *11*, 25–44 (1993).
[34] Schnörr, C.: On functionals with greyvalue-controlled smoothness terms for determining optical flow. IEEE Trans. Patt. Anal. Mach. Intell. *15*, 1074–1079 (1993).
[35] Schnörr, C.: Unique reconstruction of piecewise smooth images by minimizing strictly convex non-quadratic functionals. J. Math. Imag. Vision *4*, 189–198 (1994).
[36] Schnörr, C., Sprengel, R.: A nonlinear regularization approach to early vision. Biol. Cybernetics *72*, 141–149 (1994).
[37] Tretiak, O., Pastor, L.: Velocity estimation from image sequences with second order differential operators. In Proc. Int. Conf. Patt. Rec., pp. 16–19, Montreal, Canada, July 30–Aug. 2, 1984.
[38] Vainberg, M.: Variational method and method of monotone operators in the theory of nonlinear equations. New York: Wiley 1973.
[39] Weber, J., Malik, J.: Robust computation of optical flow in a multi-scale differential framework. In: Fourth Int. Conf. on Comp. Vision, pp. 12–20, Berlin/Germany, 1993.

Dr. C. Schnörr
Prof. R. Sprengel
Dr. B. Neumann
Universität Hamburg
FB Informatik, AB KOGS
Vogt-Köllu-Str. 30
D-22527 Hamburg
Germany

Computing Suppl. 11, 167–182 (1996)

Banach Constructor and Image Compression

W. Skarbek, Saitama

Abstract

Banach Constructor and Image Compression. The Banach constructor is defined as a concept unifying special cases of deterministic fractal modeling. The fractal compression of digital images is presented as a Banach constructor defined by a patchwork. The patchwork concept is a formal mathematical model which allowed for: a compact definition of the fractal operator, specification of a condition for its contractivity (for all v norms, $1 \leq v \leq \infty$), and formulating conditions ensuring the required fidelity of the reconstructed image. Fast fractal compression algorithm (FFC) is based on patchworks which are affine (with contrast and scaling fixed), sparse, and local. Formulas for the best fit, affine, contrast fixed transforms which perform the best fit of two digital patches, are given for v norms with $v = 1$, $v = 2$, and $v = \infty$. Experiments confirm superiority of quadratic norm at quality-time tradeoff.

Key words: Contractive operators, image compression, fractals.

1. Introduction

In the last 15 years, the fractal geometry has shown its potential in diverse areas of science and technology.

Saving the secondary storage and the transmission time, the image compression is of utmost importance in contemporary computer systems.

After Barnsley's book [2] (1988) where he claimed finding an effective way for fractal compression of images, using Iterative Function Systems, there was a great interest in the subject.

The first paper describing an automatic fractal compression method was published by Jacquin in 1989 (a conference paper) and in an extended form in 1992 ([8]).

Jacquin's method is based on the construction for the given image f a set of local transforms which work on certain image blocks producing another image blocks. As a result a (hopefully) contractive mapping F in space of images is obtained. Providing that $F(f)$ is near to f the unique fixed point of F is near to f too. Once F is known for a decoder, it makes the reconstruction of the original image f starting with arbitrarily chosen image f_0, and generating iterative sequence of images $f_{i+1} \doteq F(f_i)$. After 10–15 iterations the produced image usually gives a good approximation of f.

A local transform $T \in \mathscr{A}$ of the image block $f_{B'}$ is an affine transform acting separately in the geometric domain (affine mapping $G: B' \rightarrow B$) and in the intensity range (affine mapping $\varphi: \mathscr{R} \rightarrow \mathscr{R}$), where $\mathscr{R}$ denotes the set of real numbers. As a result, we get image block f_B' which is an approximation of the original image block f_B:

$$f_B'(X) \doteq \varphi(f(G^{-1}(X))) \quad \text{for any } X \in B.$$

The mapping φ is searched in the form:

$$\varphi(x) \doteq c \cdot x + o \quad \text{for any } x \in \mathscr{R},$$

where c is the contrasting coefficient (the contrast) and o is the brightness shifting (the offset).

By freezing the contrast c we obtain a subclass of local operators denoted by $\mathscr{A}_c$.

A local transform is optimized over parameters G, c, o in such a way that a norm $\| f_B - f_B' \|$ is minimal. This expression is further minimized over the set of admissible image domains B'.

All authors ([3, 4, 7, 8]) are solving this optimization problem over all admissible quadruples B', G, c, o, but with different meaning of admissibility for B', G, c.

For instance, Barnsley and Hurd put the fixed contrast $c = 0.75$, Jacquin chooses $c \in \{0.7, 0.8, 0.9, 1.0\}$, Jacobs et al. solve full least square problem for (c, o), but they quantize the contrast c to ten levels in the logarithmic scale. Beaumont uses a normalization principle to calculate c, and then makes optimization over o.

All authors use square domains B' spaced evenly horizontally and vertically by the length of the target domains B which are scaled down by two. Jacobs et al. admit square domains not only parallel to image sides, but rotated by 45° too.

Fixing the scaling factor to 0.5, reduces admissible geometric mappings to two forms

$$G(Y) \doteq \frac{1}{2}\begin{bmatrix} \pm 1 & 0 \\ 0 & \pm 1 \end{bmatrix} Y + \begin{bmatrix} u \\ v \end{bmatrix} \quad \text{for } Y \in B'.$$

or

$$G(Y) \doteq \frac{1}{2}\begin{bmatrix} 0 & \pm 1 \\ \pm 1 & 0 \end{bmatrix} Y + \begin{bmatrix} u \\ v \end{bmatrix} \quad \text{for } Y \in B'.$$

Hence the mapping G can be represented by a triple (g, u, v), where g identifies one of the eight matrices, and (u, v) are the coordinates of the translation vector.

The set of local transforms found in the fractal encoding stage is called LIFS (Local Iterative Function System) according Barnsley and Hurd ([3]), or PIFS (Partitioned Iterative Function System) according Fisher, Jacobs, and Boss ([7, 10]).

Explaining the mathematical background of the method, all authors refer to the IFS system of Hutchinson ([9]) and the collage theorem of Barnsley ([2]) which were developed for fractal sets. The *key fact* of this theory is *the fixed point theorem* for contractive mappings in complete metric spaces. This theorem was stated and proved by Stefan Banach in his PhD thesis written in 1922. The copy of his work was included in *The collected papers of Stefan Banach* ([1]). People working in the functional analysis or in the topology always refer to Banach's name while using his theorem.

In Section 2 we unify many special applications of the deterministic fractal modelling (such as shape representation, curve modelling, signal, and image compression) using the concept of *the Banach constructor*.

The fractal compression of digital images is presented as a special Banach constructor defined by a patchwork.

The notion of *patchwork* is used here as a formal model which gives a mathematical framework to define precisely needed terms and to prove facts useful in design of correct and efficient fractal compression algorithms.

In Section 3, we define a Banach space of images, notions of patch and patchwork, and introduce a compact definition of the fractal operator which is assigned to the given patchwork. Formally the fractal operator is the value of Banach constructor applied to the original image. We derive also sufficient conditions ensuring the required fidelity of the reconstructed image.

We specify a condition for the contractivity of the fractal operator. This condition is valid for all norms v, $1 \leq v \leq \infty$. In the literature ([3]) only the case $v = \infty$ is found.

Section 3 also includes formulas of affine transforms (with the fixed contrast) which perform the best fit of two digital patches, for norms with $v = 1$, $v = 2$, and $v = \infty$. Many conducted experiments confirm the superiority of the quadratic norm at quality-time tradeoff.

Presented in the Section 4, a Fast Fractal Compression (FFC) algorithm is based on patchworks which are affine (with the contrast and the scaling fixed), sparse, and local. While the known fractal compression schemes (Jacquin, Jacobs et al., Barnsley) require encoding time 100–1000 greater than decoding time, FFC gives high quality images with this ratio not exceeding 10.

The experimental section shows that, using local patchworks combined with proper organization of patch matching process, makes significant acceleration of the standard scheme, while the high quality of the lossy image compression is preserved.

2. Banach Constructor

The methodology of fractal compression of objects from some class (such as shapes, contours, signals, images) consists of the following elements:

1. Representation of objects in a metric space $\mathcal{F}$ with a distance function ρ. $\mathcal{F}$ must be *complete*, i.e. all Cauchy sequences should have limits in $\mathcal{F}$;
2. Construction for each object $f \in \mathcal{F}$ an operator $F: \mathcal{F} \to \mathcal{F}$. F must be *contractive* with some contractive factor $\alpha(F) \in \mathcal{R}$, i.e. satisfying

$$\rho(F(g), F(h)) \leq \alpha(F) \cdot \rho(g, h) \quad \text{for any } g, h \in \mathcal{F}.$$

We demand also from F to be nearly invariant in f, i.e. $\rho(f, F(f))$ should be small;

3. Binary encoding of the operator F, i.e. assignment bits to components of F;
4. Decoding of the operator F, i.e. inverting the encoding operation what gives components of the operator F;
5. Reconstruction of the original object f from an approximation of the object $\phi(F)$ which is the unique fixed point of F. This approximation is obtained by the iterative procedure:

$$f_0 = \text{any initial object, } f_{i+1} \doteq F(f_i) \text{ for } i \geq 0.$$

In practice it is enough to make about 10 iterations in order to get a good approximation of the object f.

In the above methodology, the first stage is performed by the system designer. The second and the third stages are executed by the fractal coder and the last two by the fractal decoder.

Correctness of the fractal compression methodology directly follows from the *Banach fixed point theorem* ([1, 5]):

Theorem 1. *Let* $F: V \to V$ *be a contractive operator in a complete metric space* $\mathcal{F}$ *with the distance function* ρ *and the contractivity factor* α. *Then*

1. *For any* v, *the sequence* $v_1 = v$, $v_{i+1} = F(v_i)$, $i \geq 1$ *has a limit, say* a_v:
2. a_v *is the fixed point of* F, *i.e.* $F(a_v) = a_v$;
3. F *has only one fixed point, say* v^*;
4. *For any* u: $a_u = v^*$;
5. *For any* $\varepsilon > 0$:

$$\|v - F(v)\| \leq (1 - \alpha)\varepsilon \Rightarrow \|v - v^*\| \leq \varepsilon.$$

In compression problems, the construction of the operator F for the given object f must be an effective procedure over the given class of objects. We need here a new unifying notion which makes proper modelling of this situation.

Let $\mathcal{F}$ be a complete metric space with the distance function ρ. We denote by $\mathcal{F}^c$ the set of all contractive mappings in $\mathcal{F}$. For any $F \in \mathcal{F}^c$ its contractive factor is written as $\alpha(F)$ and its fixed point as $\phi(F)$.

Any mapping $\Gamma: \mathcal{F} \to \mathcal{F}^c$ is called the *Banach constructor in $\mathcal{F}$*.

We say that the given Banach constructor Γ is *approximating* the given $f \in \mathcal{F}$ with the fidelity $1/\varepsilon$ if and only if

$$\rho(f, \phi(\Gamma(f))) \leq \varepsilon.$$

From the Banach fixed point theorem we get the sufficient condition to provide enough approximation fidelity by the Banach constructor:

Lemma 1. *Let Γ be a Banach constructor in $\mathcal{F}$, $f \in \mathcal{F}$, $\varepsilon > 0$. If*

$$\rho(f, \Gamma(f)(f)) \leq (1 - \alpha(\Gamma(f))) \cdot \varepsilon$$

then Γ is approximating f with the fidelity $1/\varepsilon$.   □

We say that the given operator $F \in \mathcal{F}^c$ is *invariant in f with the accuracy $1/\varepsilon$* if and only if

$$\rho(f, F(f)) \leq \varepsilon.$$

As an immediate consequence of this definition and the above lemma we get:

Corollary 1. *For any Banach constructor Γ, $f \in \mathcal{F}$, $\varepsilon > 0$, and $\delta = (1 - \alpha(\Gamma(f)))\varepsilon$, we have: if the operator $\Gamma(f)$ is invariant in f with the accuracy $1/\delta$, then Γ approximates f with the fidelity $1/\varepsilon$.*

We could conclude from this corollary that making contractive factor small, we can relax the accuracy of invariant condition. However, in practice of image compression, too high reduction of the contractive factor causes significant decrease of approximation fidelity.

3. Patch, Patchwork and Fractal Operator

Digital gray scale image with m columns and n rows with L quantization levels is represented by a 2D table of integers:

$$G = [g_{ji}], 0 \leq i < m, 0 \leq j < n, 0 \leq g_{ji} < L.$$

Fractal approach deals with analog images which are identified as real valued functions defined on rectangles of the form:

$$R_{m,n} \doteq \{(x,y)|0 \le x < m, 0 \le y < n\}.$$

We are embedding digital images into analog ones by a simple construction:

$$f_G(x,y) \doteq g_{ji} \text{ if } i \le x < i+1, j \le y < j+1.$$

A distance between two analog images f and g is measured by the v norm $\|\cdot\|$ of their difference $f - g$:

$$\|f - g\|_{v,R_{R,m}} \doteq \left(\frac{1}{mn}\int_{R_{m,n}} |f(X) - g(X)|^v \, dX\right)^{1/v},$$

where $1 \le v < \infty$.

For $v = \infty$, we define the supremum norm:

$$\|f - g\|_{\infty,R_{m,n}} \doteq \sup_{X \in R_{m,n}} |f(X) - g(X)|.$$

We consider analog images as elements of mathematical space $I_{m,n,v}$, i.e. the set of all functions defined on the rectangle $R_{m,n}$ which are absolutely v power integrable:

$$I_{m,n,v} \doteq \{f | f: R_{m,n} \to \mathcal{R}, \|f\|_{v,R_{m,n}} < \infty\}$$

$I_{m,n,v}$ is an example of a Banach space ([1, 5]).

3.1. Patches

Fractal compression works on pieces of the image which are here called patches. Patch replaces the less general notion of image block.

For the theoretical convenience, we put value zero outside of a domain of the patch and demand for the domain to be measurable set in the Lebesgue sense, denoted here as the class $\mathcal{L}_{m,n}$. However, zeros can occur in the domain of the patch. The domain is in general algorithmically elaborated part of $R_{m,n}$ and it has only vague relation to the mathematical support of the patch:

$$supp(f) = \{x \in R_{m,n} | f(x) \ne 0\}.$$

We say that $h \in I_{m,n,v}$ is a patch with the domain $B \subset R_{m,n}$ iff $supp(h) \subset B$.

In practice we use patches with domains as: squares, rectangles, or triangles.

A patch, in this application, can be induced from the original image by the restriction to a chosen domain or it can be obtained by a transformation from an induced patch.

Given $f \in I_{m,n,v}$ and $B \in \mathcal{L}_{m,n}$ we define *induced patch* as follows:

$$f_B(x) \doteq \begin{cases} f(x) & \text{if } x \in B, \\ 0 & \text{otherwise}. \end{cases}$$

The norm of a patch should be modified in order to consider less area of patch domain. If a patch h has a domain B, then its norm is:

$$\|h\|_{v,B} \doteq \left(\frac{1}{|B|}\int_B |h(X)|^v \, dX\right)^{1/v},$$

where $|B|$ = the Lebesgue measure of B and $1 \le v < \infty$.

For $v = \infty$, we have the supremum norm:

$$\|h\|_{\infty,B} \doteq \sup_{X \in B} |h(X)|.$$

In fractal compression we restrict ourselves to finite collections of sets from $\mathcal{L}_{m,n}$. They usually create a set cover of $R_{m,n}$ or a partition of $R_{m,n}$.

Any finite set collection $\gamma \subset \mathcal{L}_{m,n}$ determines for the given image f the *set of all induced patches*:

$$P_\gamma(f) \doteq \{f_B : B \in \gamma\}$$

and the *induced function*:

$$f_\gamma \doteq \sum_{B \in \gamma} f_B.$$

Three observations on partition patches will be further useful:

Fact 1. *If γ is a partition of $R_{m,n}$, then $f = f_\gamma$.*

Fact 2. *If γ is a partition of $R_{m,n}$, then for any $f \in I_{m,n,v}$ and $1 \le v < \infty$ we have*

$$\|f\|_{v,R_{m,n}}^v = \sum_{B \in \gamma} \lambda_B \|f_B\|_{v,B}^v$$

where $\lambda_B \doteq \dfrac{|B|}{mn}$.

Fact 3. *If γ is a partition of $R_{m,n}$, then for any $f \in I_{m,n,\infty}$ we have*

$$\|f\|_{\infty,R_{m,n}} = \max_{B \in \gamma} \|f_B\|_{\infty,B}.$$

We are going to fit two patches using transforms from a certain class. The *target patch* will be approximated by the result of application of a transform to the *source patch*. In practice we choose well parametrized classes of transforms such as affine transforms and look for such parameters that minimum approximation error is achieved. The search is restricted to such transforms which modify patches with domains in a given set cover γ to patches with domains in a certain set partition π of $R_{m,n}$.

From a computational point of view it is efficient to consider only *separable patch transforms*, i.e. where the resulting patch g has the domain $B = G(B')$, B' is the domain of the source patch h, G is a geometric transform in $R_{m,n}$ and for certain intensity mapping $\varphi\colon \mathcal{R} \to \mathcal{R}$

$$g(X) = \varphi(h(G^{-1}(X))\quad \text{for any } X \in B.$$

The analytical form for G and φ, used in practice, is given in the Introduction.

If a patch transform T_B from a class $\mathcal{T}$ approximates a target patch f_B from a source patch $f_{B'}$, then the approximation error is measured by $\|f_B - T_B(f_{B'})\|_{v,B}$.

If at matching two patches, the matching error achieves its minimum in the class $\mathcal{T}$, then the patch transform T is called *the best fit transform*.

3.2. Patchwork

For the given image $f \in I_{m,n,v}$ a *patchwork* is defined by specifying:

- set partition π of $R_{m,n}$;
- set cover γ of $R_{m,n}$;
- class of patch operators $\mathcal{T}$;
- patch matching function $\mu_0\colon \pi \to \gamma$;
- patch fitting function μ_1 which for any $B \in \pi$ gives a patch operator $T \in \mathcal{T}$ approximating patch f_B by patch $T(f_{\mu_0(B)})$, i.e.

$$T = \mu_1(f_B, f_{\mu_0(B)}).$$

Hence a patchwork for the given image f, to be built, requires decomposing of f into set of induced patches which are next matched to other induced patches. Matching is followed by a fitting using a patch transforms within a class $\mathcal{T}$.

Further facts will show that for image compression and good quality reconstruction, the fitting should be optimal (least norm fitting) and the best matching should provide the minimum of best fit error.

For compression efficiency we should try to reduce the cardinality of the partition π. It is usually image dependent process.

In practical circumstances only set cover γ and class of operators $\mathcal{T}$ are not dependent on image f.

More formally a patchwork Π for the image f is given by the 8-tuple:

$$\Pi = \Pi(f) \doteq (m, n, v, \pi, \gamma, \mathcal{T}, \mu_0, \mu_1).$$

For components of the patchwork Π, we use the following notation, dropping the dependence on f: $Xres_\Pi \doteq m$; $Yres_\Pi \doteq n$; $Norm_\Pi \doteq v$; $Image_\Pi \doteq f$; $Parti_\Pi \doteq \gamma$; $Cover_\Pi \doteq \gamma$; $Trans_\Pi \doteq \mathcal{T}$; $Match_\Pi \doteq \mu_0$; $Fit_\Pi \doteq \mu_1$.

Local patch operators at fitting, work on induced patches of original image f. However they can be applied on induced patches of any other function g. Combining results of all local operators we get a function defined on the whole rectangle $R_{m,n}$. This global operator is called the patchwork operator or traditionally the fractal operator.

Formally, the *patchwork operator* (*fractal operator*) $F_{\Pi(f)}$, determined by patchwork $\Pi(f)$, is defined for any $g \in I_{m,n,v}$ as follows:

$$F_{\Pi(f)}(g) \doteq \sum_{B \in \pi} T_B(g_{\mu_0(B)}),$$

where

$$\pi = Parti_\Pi, \quad T_B = \mu_1(f_B, f_{\mu_0(B)}).$$

Now we are ready to define the Banach constructor Γ in $I_{m,n,v}$:

$$\Gamma(f) \doteq F_{\Pi(f)}.$$

The Banach constructor Γ is modelling a global behaviour of fractal compression algorithms. If it approximates the given image with high fidelity, then we expect the high quality of the reconstructed images.

3.3. Contractivity and Fidelity Conditions

A compact condition for the contractivity of fractal operators is presented here for the affine class of patch transform $\mathcal{A}$.

Let $\pi = Parti_\Pi$ be the patchwork partition with p elements. Suppose that the patch transform for the i-th patch is defined by affine geometric transform G_i with the determinant of its linear part denoted by $det(G_i)$, $i = 1, \ldots, p$. The contrasting coefficient of the i-th transform is c_i.

We need a new function ρ, reflecting a degree of overlapping in the set cover $\gamma = Cover_\Pi$ defined as follows:

$$\rho(X) \doteq |\{B \in \gamma | X \in B\}|.$$

Theorem 2. *The sufficient condition for the fractal operator F_Π to be contractive is*:

$$\max_i |c_i| \left(\max_i |det(G_i)| \max_{X \in R_{m,n}} \rho(X) \right)^{1/v} < 1.$$

If $v = \infty$ this condition simplifies to: $max_i |c_i| < 1$.

It is not a suprise that to get nearness of f to $F_\Pi(f)$, it is enough to provide the good approximation of transformed source patches to target patches. Suppose that i-th patch is f_i and its approximation is f_i', $i = 1, \ldots, p$. Then

Theorem 3. *Conditions for reconstructed image fidelity*

Let the fractal operator F_Π be contractive with the contractivity coefficient α and the fixed point f^*. Then for $1 \leq v \leq \infty$, we have for arbitrary $\varepsilon > 0$:

$$\max_i \|f_i - f_i'\|_{v, B_i} < (1 - \alpha)\varepsilon \Rightarrow \|f - f^*\|_{v, R_{m,n}} < \varepsilon.$$

i.e. if all local approximation errors are less than $(1 - \alpha)\varepsilon$, then the Banach constructor Γ is approximating f with the fidelity greater than $1/\varepsilon$.

3.4. Best Fit of Digital Patches

Let $B \in \pi$ be a target patch domain discretized to K pixels. Then the source patch domain $\mu_0(B)$ can be discretized to K pixels too. Suppose that vectors

$$t = (t_1, \ldots, t_K) \text{ and } s = (s_1, \ldots, s_K)$$

represent pixel values in the target and source patch domains, respectively. Then the digital version of best fit for affine patch transform means searching for contrasting coefficient c and offset o such that

$$\sum_i |t_i - cs_i - o|^v \quad \text{is minimum}.$$

For $v = \infty$ this condition must be replaced by:

$$\max_i |t_i - cs_i - o| \quad \text{is minimum}.$$

This minimization problem can be solved very efficiently if we fix c (i.e. $\mathcal{T} = \mathcal{A}_c$) and consider three most popular special cases for norms $v = 1, 2, \infty$.

Theorem 4. *Formulas for the best fit offsets*

Given t, s and c as above. The optimal offset o is given for the following norms:

$v = 1$: $o = \text{median}_i |t_i - cs_i|$;
$v = 2$: $o = \bar{t} - c\bar{s}$, where $\bar{v}$ denotes averaging of vector v;
$v = \infty$:

$$o = \frac{\max_i |t_i - cs_i| + \min_i |t_i - cs_i|}{2}.$$

4. FFC Algorithm

In order to define efficiently a Banach constructor, we define here some conditions on the set partition π and the set cover γ.

Let δ be any distance function between sets (for instance Hausdorff metrics). Let $diam(B)$ be the diameter of the set B.

For the given patchwork Π and $B \in \pi$ we define *patchwork neighbourhood of B of radius w* as follows:

$$N_\Pi(B, w) \doteq \{B' \in \gamma \mid \delta(B, B') < w \cdot diam(B)\}.$$

The patchwork set cover γ is said to be *sparse* if the following condition is true for any $B_1 \neq B_2 \in \gamma$:

$$diam(B_1) = diam(B_2) = d \Rightarrow \delta(B_1, B_2) \geq \frac{d}{2}.$$

The patchwork Π is *local with radius w* if for any $B \in \pi = Parti_\Pi$

$$\mu_0(B) \in N_\Pi(B, w).$$

We say that the patchwork Π is *feasible* with radius w if and only if

1. $\gamma = Cover_\Pi$ is sparse;
2. for any $B \in \pi = Parti_\Pi$: B is scaled by two, affine copy of the domain $\mu_0(B)$;
3. Π is local with radius w.

A simple examples of feasible patchworks is the *uniform patchwork* with squares of size $k \times k$: π is the tiling of $R_{m,n}$ by squares of size $k \times k$, γ consists of all squares of size $2k \times 2k$ spaced evenly by k.

More compression efficiency we get from nonuniform patchworks. We build them in a top-down fashion starting from squares of size from $k_{max} \times k_{max}$. The minimum squares are of size $k_{min} \times k_{min}$. We assume that $k_{max} = 2^d k_{min}$, γ consists of all squares of size $2k \times 2k$ spaced by k, for $k_{min} \leq k \leq k_{max}$.

By $\mathcal{N}_k(B, w)$ we denote the set of all squares in γ of size k which are in $N_\Pi(B, w)$.

To preserve the space coherence of encoded parameters we visit subsquares of the given square using the Hilbert scan (implemented according Goldshlager's algorithm [6]). Beginning from the top left square corner, going clockwise, corners are identified by symbols X, Y, Z, U. The order of subsquares is called an *orientation*. To provide the continuity of scan between maximum size squares, the initial orientation should be X, U, Z, Y.

Fractal compression algorithm depends on some input tuning parameters: the fixed contrast c, neighbourhood radius w, maximum block size k_{max}, minimum block size k_{min} ($k_{max} = 2^d k_{min}$), and the best fitting threshold ε.

Patchwork construction is implemented in consecutive blocks of maximum size $k_{max} \times k_{max}$ by a recursive procedure *FFC*:

Algorithm. *Fast Fractal Compression*

$FFC_k(f_B, x, y, z, u)$ {

  $e_0 = \infty;$

  for each $B' \in \mathcal{N}_{2k}(B, w)$ {

    $e_1 = \infty;$

    for each affine $G: B' \to B$ {

      $o = \bar{f}_B - c \cdot \overline{f_{B'} \circ G^{-1}};$

      $f'_B = c \cdot f_{B'} \circ G^{-1} + o;$

      $e = \| f_B - f'_B \|_{2,B};$

      if $e < e_1$ {$G_{min} = G; o_{min} = o; e_1 = e;$ }

    }

    if $e_1 < e_0$ {$B'_{min} = B'; e_0 = e_1;$ }

  }

  if $e_0 \leq \varepsilon$ **or** $k = k_{min}$ {

    if $k \neq k_{min}$ { Emit subdivision bit $b = 0;$ }

    Emit bits for normalized location differences $\frac{4x}{k}, \frac{4y}{k}$ between B'_{min} in and B;

    Emit bits of $code(G_{min})$;

    Emit bits of offset o_{min};

    return;

    }

  Divide the domain B into subdomains B_x, B_y, B_z, B_u;

  Emit subdivision bit $b = 1$;

  $FFC_{k/2}(f_{B_x}, x, u, z, y);$

  $FFC_{k/2}(f_{B_y}, x, y, z, u);$

  $FFC_{k/2}(f_{B_z}, x, y, z, u);$

  $FFC_{k/2}(f_{B_u}, z, y, x, u);$

}

In the actual implementation, domains $B' \in \mathcal{N}_{2k}(B, w)$ are accessed along a spiral centered in B.

The affine mapping G can be divided into two parts: G_1—the scaling by 0.5, and G_2—an isometric part (a rotation with optional mirror reflection). The scaling mapping can be precomputed for the whole image and kept as a separate image (provided k_{min} is even). The isometric part can be factored out of the B' loop if we only precompute for the given B all possible preimages $G_2^{-1}(B)$ (for square domains there are eight such preimages). This trick makes significant acceleration of the scheme for $w > 1$.

Each coding component is sent to separate bit stream, hence we have five bit streams. It enables their efficient entropy coding.

Small improvement we can get by cancelling parameters y, z in FFC, since they can be uniquely determined from the input corner x and the output corner u.

5. Experimental Results

In this section some experimental results[1] for gray scale images and uniform patchworks are presented. They lead to conclusions that the best norm is for $v = 2$ with the best fit threshold $\varepsilon = 4$. The best block size for uniform patchworks is $k = 4$. The good contrasting coefficient $c \in [0.5, 0.7]$. Satisfactory feasible patchworks have radius $w = 2$ or $w = 3$.

Results for colour images show increase of compression efficiency by more than twice. Using nonuniform patchworks gives further increase of compression efficiency by a factor equal at least two.

In tables below the following measures are used:

- **mse**: mean squared error;
- **msnr**: maximum signal to noise ratio;
- **M%**: threshold matching percentage;
- **e**: total entropy;
- **r**: compression ratio estimate;
- **t/τ_{10}**: encoding time divided by time of decoder's 10 iterations;
- **eye**: subjective quality evaluation.

Total entropy formula is:

$$e = \text{entropy of } (xSource - xTarget)$$
$$+ \text{ entropy of } (ySource - yTarget)$$
$$+ \text{ entropy of } (CurrOffset - PrevOffset)$$
$$+ \text{ conditional entropy of } (GeometricCode)$$

The conditional entropy is computed relatively to the context obtained from four neighbourhood target domains placed in directions W, WN, N, NE.

Compression ratio r is estimated by dividing block volume in bits by total entropy e.

Subjective quality evaluation codes are: E(xcellent, V(ery good, G(ood, P(oor, B(ad.

5.1. Changing the Block Size

For small target patches ($k = 3, 4$) and corresponding target patches ($l = 9, 8$) the image quality is very good. Increasing k and l gives higher compression r with decreasing of image fidelity. Good tradeoff is achieved for uniform patchworks at $k = 4, l = 8$—see Table 1.

---

[1] The program implementing FFC is available by *anonymous fip* at site *wars.ipipan.waw.pl* in the catalog */pub/fractal* in the file *appendix.tex*.

Table 1. Results for a human face image. Checking for block size impact. Target block is of size $k \times k$; source block is of size $l \times l$; $v = 2$, $\varepsilon = 4$, $c = 0.5$, $w = 3$

k	l	mse	msnr	M%	e	r	t/τ_{10}	eye
3	9	9.87	38.19	94.8	9.27	7.77	4.0	E
4	8	14.96	36.38	86.9	9.29	13.77	6.0	V
4	16	16.07	36.07	79.4	10.77	11.88	8.8	G
5	25	1173.58	17.44	61.0	11.49	17.40	12.8	B
6	12	24.80	34.19	66.5	10.52	27.37	12.0	P
6	18	24.60	34.22	62.0	11.09	25.97	13.4	P
6	36	8756.92	8.71	42.1	11.89	24.22	18.0	B
8	16	37.24	32.42	49.1	11.02	46.44	15.2	B
9	27	43.80	31.72	39.0	11.44	56.66	17.2	B
16	32	107.37	27.82	14.5	11.44	179.08	21.8	B

5.2. Changing the Neighbourhood Radius

Table 2. Results for a human face image. Checking for searching window impact. $v = 2$, $\varepsilon = 4$, $c = 0.5$, $k = 4$, $l = 8$

w	mse	msnr	M%	e	r	t/τ_{10}	eye
0	26.01	33.98	62.3	6.27	20.42	1.0	P
1	16.43	35.98	79.8	8.28	15.47	2.4	G
2	15.49	36.23	84.5	8.93	14.34	4.2	V
3	14.96	36.38	86.9	9.29	13.77	6.0	V
4	14.64	36.47	88.5	9.54	13.42	8.2	V
5	14.45	36.53	89.7	9.68	13.23	10.8	V
6	14.30	36.58	90.5	9.80	13.06	13.4	V
∞	12.83	37.05	95.5	10.66	12.00	134.4	V

For small searching windows, encoding time is very short (it is comparable with the decoding time), but the quality is poor. At $w = 2, 3$ the quality is already very good. Further increase of w improves slightly the image fidelity in cost of significant increase of encoding time (cf. Table 2).

The basic Jacquin scheme ([8]) corresponds to uniform patchworks at $w = \infty$. Comparing to the local patchwork of radius $w = 3$ we have improvement of image fidelity by less than one decibel but in time at least 20 times longer.

While the known fractal compression schemes require encoding time 100–1000 greater than decoding time, FFC gives high quality images of natural scenes with this ratio not exceeding 10. In the actual implementation on PC 486 this time varies from 10–30 seconds.

5.3. Cutting Bits in Offset Coefficient

Cutting of $q = 1, 2$ bits in the brightness offset o causes no special harm for the image quality and increases compression efficiency by about 10% (see Table 3).

Table 3. Results for a colon, medical, endoscopic image. Checking for bit cutting impact. $v = 2$, $\varepsilon = 4$, $w = 2$, $c = 0.6$, $k = 4$, $l = 8$

q	mse	msnr	M%	e	r	t/τ_{10}	eye
0	42.68	31.83	83.0	8.86	14.44	4.0	V
1	44.69	31.63	83.0	7.90	16.20	4.0	V
2	57.74	30.52	83.0	6.95	18.41	4.0	G
3	119.01	27.37	83.0	6.12	20.90	4.0	P
4	397.28	22.14	83.0	5.47	23.40	4.0	B
5	1180.74	17.41	83.0	4.95	25.88	4.0	B

5.4. Changing the Best Fit Threshold

Fixing small threshold ε for the local approximation error gives very good quality but it needs longer encoding time.

For the square norm ($v = 2$) the good choice is $\varepsilon = 4$, giving reasonable time and still very good quality (Table 4).

Table 4. Results for a human face image. Checking for best fit threshold impact. $v = 2$, $w = 2$, $c = 0.6$, $k = 4$, $l = 8$

ε	mse	msnr	M%	e	r	t/τ_{10}	eye
1	12.48	37.17	0.0	11.11	11.52	15.4	V
2	12.51	37.16	13.5	11.07	11.56	14.0	V
3	13.25	36.91	59.3	10.43	12.27	8.0	V
4	14.75	36.44	85.1	9.02	14.19	4.0	V
5	16.31	36.01	93.8	7.97	16.06	2.4	G
7	19.08	35.33	98.2	7.06	18.14	1.6	P
8	20.24	35.07	98.9	6.84	18.70	1.4	P
9	21.45	34.82	99.2	6.70	19.09	1.4	P
10	22.31	34.65	99.4	6.60	19.41	1.4	P
15	25.88	34.00	99.9	6.35	20.16	1.0	P
20	27.73	33.70	100.0	6.26	20.45	1.2	P

5.5. Changing the Contrasting Coefficient

Table 5. Results for a highway image. Checking for contrasting coefficient impact. $v = 2$, $\varepsilon = 4$, $w = 2$, $k = 4$, $l = 8$

c	mse	msnr	M%	e	r	t/τ_{10}	eye
0.1	18.74	35.40	88.6	6.20	20.64	2.8	B
0.2	16.09	36.07	90.7	6.30	20.31	2.6	B
0.3	13.91	36.70	91.9	6.23	20.53	2.4	P
0.4	12.34	37.22	92.6	6.19	20.69	2.4	P
0.5	11.26	37.61	93.0	6.20	20.63	2.4	G
0.6	10.68	37.85	93.1	6.21	20.61	2.4	G
0.7	10.55	37.90	93.1	6.27	20.43	2.4	V
0.8	11.26	37.62	92.9	6.33	20.21	2.4	G
0.9	22.96	34.52	92.6	6.41	19.98	2.4	P
0.99	273.83	23.76	92.1	6.51	19.66	2.4	B

Small contrasting coefficient gives usually bad quality images. It follows from the fact that small c causes small differences in the transformed source patch which at fitting to the target patch cannot be compensated by an optimal offset (cf. Table 5).

The value of c to close to one gives according the fidelity Theorem 3 too strict requirements for the actual approximation error. Usually these requirements cannot be satisfied what gives the fixed point of the fractal operator far away from the original image.

In some situations c close or greater than one may violate the contractivity property (see Theorem 2) and hence the fractal operator can have more than one attractors not related to the original image.

A good choice for c at uniform patchworks are values from the interval $[0.5, 0.7]$.

6. Conclusions

We have developed a patchwork theory for the fractal compression method within which contractivity and fidelity conditions were obtained.

A fast fractal compression algorithm FFC could be entirely specified in terms of patchwork notions. It gives at least one order of magnitude faster fractal coder than others known from the literature. The acceleration is based on the local patchwork design and on the proper factorization of the geometric mappings out of the algorithm most inner loop.

Nonuniform patchwork approach gives the state of art schema in which the Hilbert 1scan allows for even further increase of compression efficiency by the entropy coding.

References

[1] Banach, S.: Oeuvres, Vol. I and Vol. II. Warszawa: Polish Scientific Publishers 1978.
[2] Barnsley, M.: Fractals everywhere. San Diego: Academic Press 1988.
[3] Barnsley, M., Hurd, L.: Fractal image compression. Wellesley: AK Peters 1993.
[4] Beaumont, J.: Image data compression using fractal techniques. BT Techn. J. *9*, 93–108 (1991).
[5] Dugundi, J., Granas, A.: Fixed point theory. Warszawa: Polish Scientific Publishers 1982.
[6] Goldshlager, L.: Short algorithms for space filling curves. Software Pract. Exp. *11*, 99–100 (1981).
[7] Jacobs, E., Fisher, Y., Boss, R.: Image compression: a study of the iterated transform method. Signal Proc. *29*, 251–263 (1992).
[8] Jacquin, A.: Image coding based on a fractal theory of iterated contractive image transformations. IEEE Trans. Image Proc. *1*, 18–30 (1992).
[9] Hutchinson, J.: Fractals and self-similarity. Indiana Univ. J. Math. *30*, 713–747 (1981).
[10] Peitgen, H., Jurgens, H., Saupe, D.: Chaos and fractals. New York: Springer 1992 (Appendix written by Y. Fisher).

Dr. W. Skarbek
Laboratory for Artificial Brain Systems
The Institute of Physical
and Chemical Research (Riken)
z-1, Hirosawa, Waka-shi
Saitama 351-01
Japan
e-mail: ska@negi.riken.go.jp

Computing Suppl. 11, 183–199 (1996)

Piecewise Linear Approximation of Planar Jordan Curves and Arcs: Theory and Applications

F. Sloboda and **B. Zat'ko**, Bratislava

Abstract

Piecewise Linear Approximation of Planar Jordan Curves and Arcs: Theory and Applications. Piecewise linear approximation of planar Jordan curves and arcs based on the shortest polygonal path in a polygonally bounded compact set and on the geodesic diameter in a polygon is described and applications involving gridding techniques are highlighted.

Key words: Piecewise linear approximation, shortest path, geodesic diameter.

1. Introduction

Piecewise linear approximation of planar simple closed (Jordan) curves and arcs represents a highly actual problem both from theoretical and practical point of view. Piecewise linear continuation methods are used for approximation of numerical solutions of nonlinear systems of equations, see [2, 17]. Piecewise linear approximations are extensively used in computer graphics and computer aided geometric design in order to represent, approximate and compute curves and arcs, in image processing in order to approximate and to represent boundaries of compact sets and in line drawing image processing. Line drawing image processing has variety of applications, for example in geographic information systems for vectorization of conventional maps. The most of above-mentioned applications are related to so called gridding techniques. The gridding techniques are related to regular grids. There are three types of regular grids which are defined by square, triangular and hexagonal topological units. In the last five years the achievements in high technology allowed to produce high density scanners, plotters and monitors which have a direct impact on software required for information processing technology. The amount of information obtained by high resolution scanners set up new demands on algorithms, their effectiveness and complexity.

In this paper new results concerning planar Jordan curves and arcs approximation are described and their applications are highlighted. In [22, 23] a topological approach for approximation of planar Jordan curves and arcs is described. This approach is based on basic notions of intrinsic geometry of metric spaces and algorithms are related to basic problems in computational geometry. They are related to the shortest path problem solution in a polygonally bounded compact set and to the geodesic diameter calculation in a polygon. They allow to use gridding techniques so that they can be effectively used for representation, visualization and computation purposes.

2. Planar Jordan Curves Approximation

A planar Jordan curve is a simple closed curve, that is a curve which belongs to a parametrized path $\phi: [a, b] \to R^2$ with $a \neq b$, $\phi(a) = \phi(b)$, $\phi(s) \neq \phi(t)$ for all $a \leq s < t < b$. A rectifiable curve is a curve L belonging to a parametrized path, say $\phi: [a, b] \to R^2$ with bounded length $d(L)$. A planar Jordan curve L is called a polygonal curve, if it consists of finitely many linear pieces. By the Jordan curve theorem the polygonal curve is the boundary L of a bounded polygon P_L, where P_L is a simple connected compact set with a nonempty interior P_L^o. A simply connected compact set whose boundary is a polygonal curve is in literature called also a simple polygon.

In [22] a new approach for the length approximation of a rectifiable planar Jordan curve is described. This approach is related more to differential topology than to differential geometry and is based on the notion of the shortest path in a polygonaly bounded compact set. The length approximation of a rectifiable planar Jordan curve is related to the following [22]

Problem   (1)

Given two polygonal curves L_1 and L_2 of R^2 such that L_1 is contained in the interior of P_{L_2}, $P_{L_1} \subset P_{L_2}^o$, consider the annular compact set $G = P_{L_2} \backslash P_{L_1}^o$ between L_1 and L_2 with the boundary $\partial G = L_1 \cup L_2$. The problem is to find the shortest Jordan curve in G containing P_{L_1}, see Fig. 1.

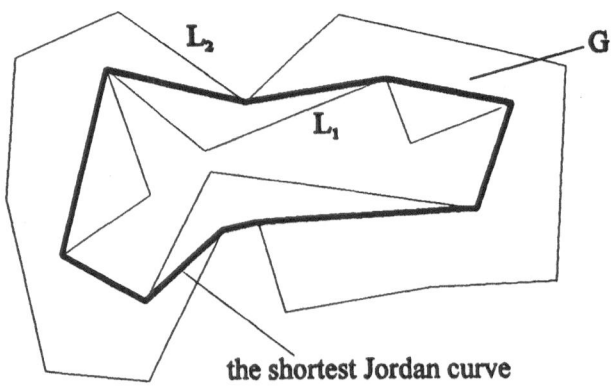

Figure 1

Problem (1) has a solution, which is unique [22]. The solution curve is a polygonal Jordan curve whose convex vertices belong to the convex vertices of L_1, and whose concave vertices belong to the concave vertices of L_2 [22]. The existence of the solution follows from general theorems on rectifiable curves and geodesics in metric spaces such as R^2, see ([1], pp. 68–82). The shortest Jordan curve in G encircling L_1 is called the minimum perimeter polygon belonging to G, see also [15, 21]. The approximation of a given rectifiable Jordan curve $\gamma\colon [0, d(\gamma)] \to R^2$ is based on the following theorem [22]:

Theorem 1. *Let $\gamma\colon [0, d(\gamma)] \to R^2$ be a rectifiable noncontractible Jordan curve in G_0 encircling $L_1^{(0)}$ in the positive sense with*

$$\Gamma \subset \cdots \subset G_{i+1} \subset G_i \subset \cdots \subset G_0, \; \Gamma = \bigcap_i G_i$$

where $\Gamma\colon \{\gamma(t)|t \in [0, d(\gamma)]\}$, and G_i are polygonal compact sets with nonempty interior and boundary $\partial G_i = L_1^{(i)} \cup L_2^{(i)}$, $L_1^{(i)}$, $L_2^{(i)}$ polygonal curves with $L_1^{(i)}$ lying in the interior of $L_2^{(i)}$. Further, let $p_i\colon [0, d_i] \to R^2$, $d_i = d(p_i)$ be the minimum perimeter polygon lying in G_i and encircling $L_1^{(i)}$ in the positive sense, parametrized by arclength. Then

$$\lim_i d(p_i) = d(\gamma),$$

(see Fig. 2).

The geometric properties of the shortest polygonal Jordan curve in a polygonally bounded compact set $G = P_{L_2} \backslash P_{L_1}^o$ encircling L_1 are described in [7]. These properties and the corresponding algorithm are based on convex hulls. The vertices of the convex hull of L_1 belong to the vertices of the shortest polygonal Jordan curve in G encircling L_1 [22]. This property and the uniqueness of the solution enable to cut G in any vertex of the convex hull of L_1 in order to obtain a polygon which has the same shortest path problem solution as the original polygonally bounded

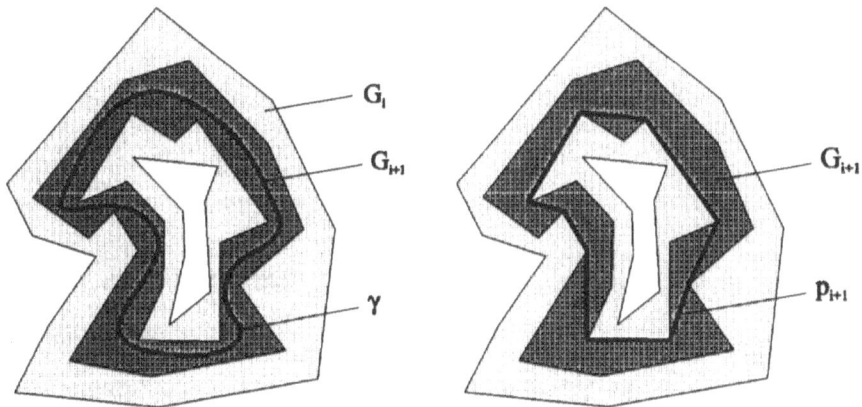

Figure 2

compact set G [24, 29]. All algorithms for the shortest path problem solution in a polygon between two specified vertices are based on partition of the polygon. The most popular partition of a polygon is the triangulation of the polygon. The triangulation problem of an n-vertex polygon P_L is to find $n-3$ nonintersecting diagonals of P_L, which partition the interior of P_L into $n-2$ triangles, whereby the diagonals of P_L are open line segments whose endpoints are vertices of P_L and that lie entirely in the interior of P_L. Being a polygon triangulated the shortest path in the polygon between two specified points can be found in linear time [13]. The triangulation of a polygon has a long, remarkable history. The first algorithm was published in 1911 [14]. Recently number of algorithms were proposed [3, 5, 8, 10, 28]. The complexity of these algorithms was improved from $O(n^2)$ [14] to $O(n\log n)$ [10], later to $O(n\log\log n)$ [28] and finally to $O(n)$ [5]. The greatest disadvantage of the triangulation is that it is in general not unique, which may cause theoretical and practical problems, for example to prove correctness of an algorithm. The only up to now published proof of the uniqueness of the shortest path in a polygon between two specified points of the polygon which is based on triangulation of the polygon [13] is not correct. Owing to the fact that triangulation is not unique, the argumentation in the proof is not right. Nevertheless, the proof of the uniqueness of the shortest path in a polygon between two specified points of the polygon is irrelevant to any partition of the polygon (see the proof of the uniqueness in [22, 23]). The vertices of the shortest path between two specified vertices in a polygon belong to the concave vertices of the polygon [23]. It means that among all vertices of the polygon only its concave vertices could belong to the shortest path in the polygon. On triangulation of a polygon all vertices of the polygon take part. Therefore the triangulation is not the most efficient procedure for the shortest path problem solution in a polygon.

Another important partition is monotone polygon partition of a polygon. A monotone polygon partition of a polygon P_L is partition of P_L into monotone polygons whose vertices belong to P_L. A polygonal path is monotone with respect to a straight line if orthogonal projections of its vertices onto the straight line are ordered the same as the vertices of the polygonal path. A polygon is monotone if it can be partitioned into two polygonal paths monotone with respect to the same straight line. The concept of monotone polygons was introduced in [12]. In [11, 10] $O(n\log n)$ algorithms for monotone polygon partition were suggested. Because triangulation of a monotone polygon can be performed in linear time [4, 10], triangulation of a polygon using monotone polygon partition algorithm may be performed in $O(n\log n)$ time [10, 16].

Triangulation and monotone polygon partition of a polygon is based on trapezoidization of the polygon. Trapezoidization is to find all x-axis parallel pseudodiagonals of the polygon. In fact the time complexity of triangulation of a polygon is given by the complexity of the trapezoidization. Being a polygon trapezoidated triangulation may be achieved in linear time [3, 8]. Usually trapezoidization requires sorting, which leads to $O(n\log n)$ complexity. In [28] it is shown that sorting can be avoided to achieve $O(n\log\log n)$ complexity. This result was improved in [5] to $O(n)$ time. These results are valuable theoretical results, but from practical point of view are of less interest [31].

In [11] it is shown that practical improvement can be obtained if trapezoidization is realized only for concave vertices of the polygon. By this modification (using sorting) trapezoidation can be realized in $O(n + r \log r)$ time [11, 16], and in $O(n \log r)$ time [8] where r is the number of concave vertices. Similarly in [3] it is shown that trapezoidation can be performed in $O(n \log s)$ time where s is the "sinuosity" of the polygon, which may be $O(n)$ but usually it is very small.

Further improvement can be achieved introducing extremal vertices of a polygon related to y-axis [25, 26]:

Definition 1. Let $v_0(x_0, y_0), \ldots, v_{n-1}(x_{n-1}, y_{n-1})$ be vertices of polygon P_L. A vertex $v_i(x_i, y_i)$ of P_L is an extremal vertex of P_L if

$$y_{i-1} < y_i > y_{i+1 (\text{mod } n)} \lor y_{i-1} > y_i < y_{i+1 (\text{mod } n)},$$

and two neighbouring vertices $v_i(x_i, y_i)$, $v_{i+1}(x_{i+1}, y_{i+1})$ of P_L with $y_i = y_{i+1}$ are extremal vertices of P_L if

$$y_{i-1} < y_i > y_{i+2 (\text{mod } n)} \lor y_{i-1} > y_i < y_{i+2 (\text{mod } n)}.$$

Extremal concave vertices related to y-axis allow to define pseudomonotone polygon partition [26]. A pseudomonotone polygon partition of polygon P_L is partition of P_L by x-axis parallel pseudodiagonals of P_L into monotone polygons. The pseudomonotone polygon partition can be performed in $O(n + r^* \log r^*)$ time where r^* is the number of extremal concave vertices of P_L [26]. Further, in [26] it has been shown that the shortest path problem in a monotone polygon between its two vertices can be solved in linear time without any partition of the polygon. As a consequence we have that the shortest path problem in a polygon may be solved in $O(n + r^* \log r^*)$ time [26].

Pseudomonotone polygon partition can be applied also on polygonally bounded compact set $G = P_{L_2} \backslash P_{L_1}^o$. For this purpose we shall consider extremal convex vertices of L_1 and extremal concave vertices of L_2. The shortest path problem solution in $G = P_{L_2} \backslash P_{L_1}^o$ can be obtained by the following algorithm:

Algorithm   (2)

Step (1): Find all extremal convex vertices of P_{L_1} and all extremal concave vertices of P_{L_2}, see Fig. 3.
Step (2): Find all x-axis parallel pseudodiagonals of G related to extremal convex vertices of P_{L_1} and extremal concave vertices of P_{L_2}, see Fig. 3.
Step (3): Eliminate all monotone polygons whose boundaries contain only vertices of L_1 or L_2, respectively; and denote the resulting polygonally bounded set by G', see Fig. 3.
Step (4): Find shortest path in each monotone polygon between two extremal vertices of G', see Fig. 3.

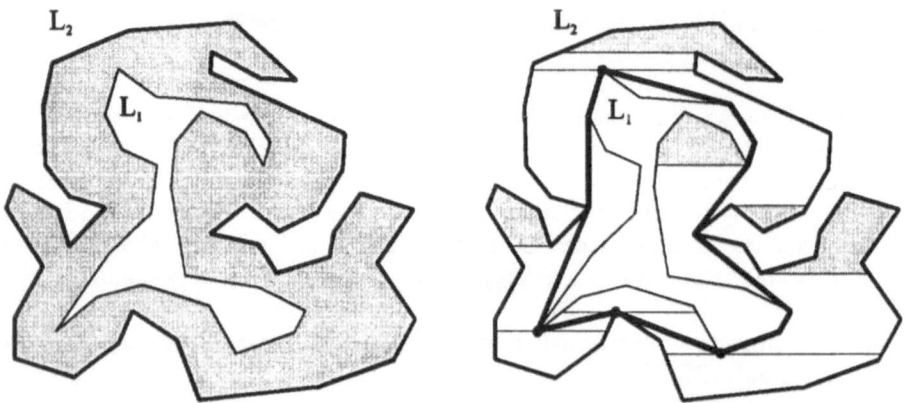

Figure 3

Step (1), Step (3) and Step (4) can be performed in linear time [26]. Step (2) requires sorting of extremal convex and concave vertices which takes $O(r^* \log r^*)$ time where r^* is the number of extremal convex and concave vertices of P_{L_1}, P_{L_2}, respectively. The x-axis parallel pseudodiagonals of G can be then found in $O(n + r^* \log r^*)$ time. In the following we will show that in the case of approximation of planar Jordan curves by a gridding technique, r^* does not depend on n which implies that for a large class from practical point of view important polygonally bounded sets the shortest path problem can be solved in linear time.

3. Planar Jordan Arcs Approximation

A planar Jordan arc is a simple curve, that is a curve which belongs to a parametrized path $\phi: [a,b] \to R^2$ with $a \neq b$, $\phi(a) \neq \phi(b)$, $\phi(s) \neq \phi(t)$ for all $a \leq s < t < b$. A rectifiable planar Jordan arc is a simple curve L belonging to a parametrized path, say $\phi: [a,b] \to R^2$ with bounded length

$$d(L) = \sup_{t_i: a = t_1 < \cdots < t_n = b} \sum_{1}^{n} \| \phi(t_i) - \phi(t_{i-1}) \| < \infty. \tag{3}$$

A planar Jordan arc L is called a polygonal arc, if it consists of finitely many linear pieces. According to (3) the length of a planar Jordan arc can be approximated by the length of a polygonal Jordan arc whose vertices are lying on the trace of the planar Jordan arc. This theoretical approach how to approximate the length of a rectifiable Jordan arc causes practical problems. Similarly as in the case of planar Jordan curves also in the case of planar Jordan arcs it was highly actual to find another approach how to define the length of a rectifiable planar Jordan arc, which has more practical applications as the classical one. In [23] such a new approach has been described. This approach is based on the notion of the geodesic diameter of a polygon and is related to the following problem:

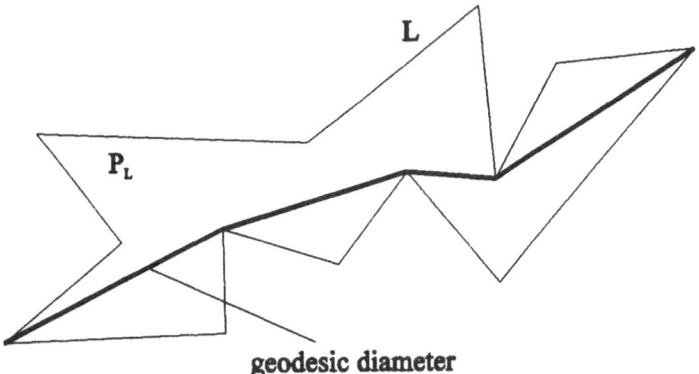

geodesic diameter

Figure 4

Problem   (4)

Given a polygonal curve L, the boundary of a polygon P_L, find a geodesic diameter of P_L, see Fig. 4.

The geodesic diameter of P_L is a shortest path of maximal length internal to P_L. Problem (4) has a solution [23], and any solution is a polygonal Jordan arc whose vertices and endpoints belong to the vertices of P_L [23]. The endpoints are convex vertices of P_L and the remaining vertices belong to the concave vertices of P_L [23].

In order to approximate a rectifiable planar Jordan arc $\gamma: [0, d(\gamma)] \to R^2$ let us consider a set of nested polygons $P_{L^{(i)}}$, all shrinking to the trace of γ. In [23] it has been shown that the sequence of geodesic diameter lengths corresponding to $P_{L^{(i)}}$ converges to the length of γ. It was necessary to exclude some pathologic possibilities related to the set of $P_{L^{(i)}}$. Therefore the following assumption has been introduced [23]:

Definition 2. A polygon P_L containing a rectifiable Jordan arc $\Gamma = \{\gamma(t) | 0 \leq t \leq d(\gamma)\}$ is called c-nondegenerate if for any vertex u of P_L there exists an $x \in \Gamma$ and a shortest path p in P_L connecting u to x of length satisfying

$$d(p) \leq c. \max_{y \in P_L} \left\{ \min_{z \in \Gamma} \| y - z \| \right\}.$$

The convergence of the corresponding geodesic diameter lengths follows then from the following theorem [23]:

Theorem 2. *Let $\gamma: [0, d(\gamma)] \to R^2$ be a rectifiable planar Jordan arc and let*

$$\Gamma \subset \cdots \subset P_{L^{(i+1)}} \subset P_{L^{(i)}} \subset \cdots \subset P_{L^{(0)}} \text{ and } \Gamma = \bigcap_i P_{L^{(i)}},$$

where $\Gamma := \{\gamma(t) | t \in [0, d(\gamma)]\}$. Let $g_i: [0, d_i^+] \to R^2$ be a geodesic diameter of $P_{L^{(i)}}$ with length $d_i^+ := d(g_i)$. If there is a $c > 0$ so that all $P_{L^{(i)}}$, $i \geq 0$ are c-nondegenerate,

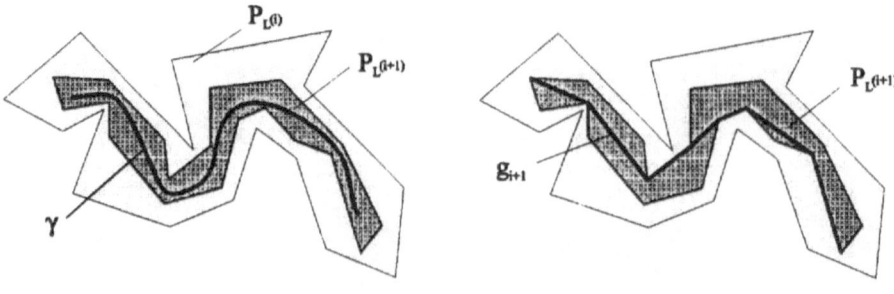

Figure 5

then the geodesic diameter lengths d_i^+ of $P_{L^{(i)}}$, $i \geq 0$ converge to $d(\gamma)$

$$\lim_i d_i^+ = d(\gamma),$$

(*see Fig.* 5).

In the following it will be shown that a large class of practical important nested polygons $P_{L^{(i)}}$ satisfying the conditions of the theorem are c-nondegenerate for all $i \neq 0$.

A geodesic diameter of a polygon is in general not unique. There are $O(n^2)$ shortest paths in an n-vertex polygon whose endpoints are convex vertices of the polygon which indicates that by a brute-force method a geodesic diameter can be calculated in $O(n^2 T(n))$ time complexity, where $T(n)$ is the time complexity for the shortest path problem solution between two specified vertices in a polygon, i.e., according to [5] it can be calculated in $O(n^3)$ time. In [4] $O(n^2)$ time algorithm was suggested. Even this result was improved to $O(n \log n)$ time complexity [27], see also [30]. The above described techniques for planar Jordan curves and arcs approximation represent the basis for gridding techniques.

4. Gridding Techniques

Regular grids are extensively used in numerical mathematics in order to solve boundary value problems of partial differential equations or to approximate numerical solutions of nonlinear systems of equations [2, 17]. In this case the gridding technique is used for piecewise linear approximation of a piecewise linear curve which approximates a smooth one. The regular grids represent also basic tools of image processing, computer graphics and computer aided geometric design. Computer aided geometric design is the representation, approximation and computation of curves, surfaces and volumes and has variety of applications. There are only three types of regular grids which are defined by identical nonoverlapping regular convex topological units, the union of which covers the whole plane R^2. These are orthogonal, triangular and hexagonal grids.

In the following the orthogonal grids will be used for planar Jordan curves and arcs representation and approximation purposes using the topological approach described in the previous sections. In order to approximate planar Jordan curves (simple closed curves) and planar Jordan arcs (simple open curves) by a gridding technique let us consider square grids on R^2. For $p = 0, 1, 2, \ldots$ and for each tuple (w_1, w_2) of integer numbers let

$$N^p_{(w_1, w_2)} = \{w \in R^2: w_i 2^{-p} \leq x_i \leq (w_i + 1)2^{-p}, i = 1, 2\}. \tag{5}$$

Let $\gamma: [a, b] \to R^2$ be a continuously differentiable planar Jordan curve. Let us denote

$$M_p = \bigcup_1^k N^p_{(w_1, w_2)}, k \geq 2,$$

where $N^p_{(w_1, w_2)} \cap \gamma \neq \varnothing$. Then there exists p such that for $p + k$, $k = 0, 1, \ldots$, the following theorem [22] holds:

Theorem 3. *Let $\gamma: [a, b] \to R^2$ be a continuously differentiable planar Jordan curve. Let p_k denote the shortest closed polygonal path (minimum perimeter polygon) corresponding to M_{p+k}, $k = 0, 1, \ldots$. Then*

$$\lim_k d(p_k) = d(\gamma),$$

(see Fig. 6).

Theorem 3 is a consequence of the general Theorem 1, and shows the convergence of the shortest polygonal path lengths to the length of γ. Theorem 3 represents a modification of the Theorem 7 [22], where γ was considered to be the boundary of a planar compact set M and only those $N^p_{(w_1, w_2)}$ were considered which have had nonempty intersection with M and which have not lain in M.

Let $\gamma: [a, b] \to R^2$ be a continuously differentiable planar Jordan curve which has the form

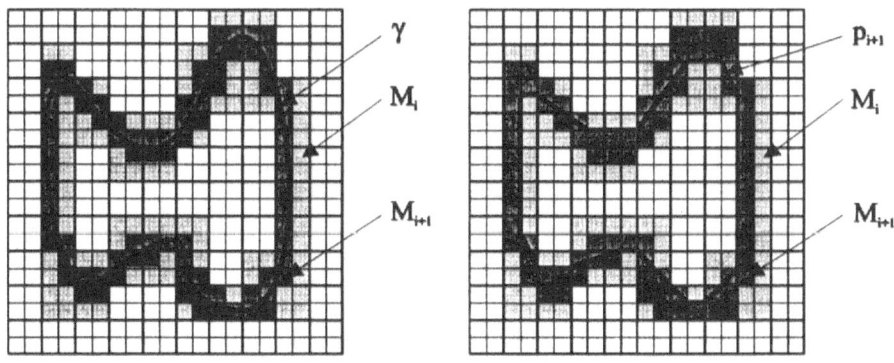

Figure 6

Table 1

Size	64 × 64	128 × 128	256 × 256	512 × 512	1024 × 1024	2048 × 2048
$|M_p|$	260	500	1148	2272	4428	8856
$|MPP|$	28	50	82	127	207	321
ER	1.5146	0.6077	0.2083	0.0997	0.0308	0.0153
θ	42.4088	30.385	17.0806	12.6619	6.3756	4.9113

$|M_p|$, $|MPP|$ denote the number of vertices of M_p, the number of the minimum perimeter polygon (MPP) vertices, respectively, ER denotes the length error in % related to the length of γ obtained by the Gaussian quadrature, and $\theta = ER \times |MPP|$ represents the measure of effectiveness of the polygonal approximation

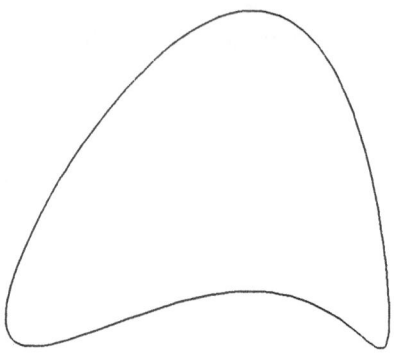

Figure 7

$$x(t) = \sin(t)(3\sin(t) + 2.5)$$
$$y(t) = \cos(t)(\cos(t) + 3.5), \qquad t \in [0, 2\pi]. \tag{6}$$

The approximation of (6) by the shortest polygonal curve (minimum perimeter polygon) on different square grids is shown in Table 1 in corresponding columns.

The resulting shortest polygonal Jordan curve which corresponds to the 512 × 512 square grid points resolution is shown on Fig. 7.

Let $\gamma: [a, b] \rightarrow R^2$ be a continuously differentiable planar Jordan arc. Let us denote

$$M_p = \bigcup_1^k N_{(w_1, w_2)}^p, \, k \geq 2,$$

where $N_{(w_1, w_2)}^p \cap \gamma \neq \varnothing$ and $N_{(w_1, w_2)}^p$ is defined by (5). Then there exists p such that for $p + k$, $k = 0, 1, \ldots$, the following theorem [23] holds:

Theorem 4. *Let* $\gamma: [a, b] \rightarrow R^2$ *be a continuously differentiable planar Jordan arc. Let* g_k *denote the geodesic diameter corresponding to* M_{p+k}, $k = 0, 1, \ldots$. *Then*

$$\lim_k d(g_k) = d(\gamma),$$

(*see Fig. 8*).

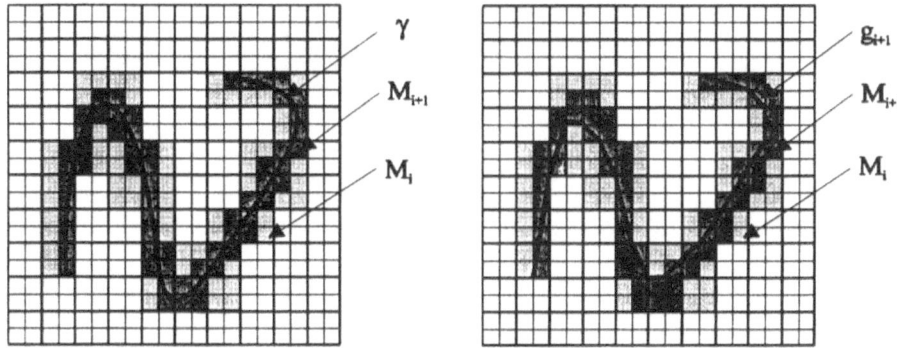

Figure 8

Table 2

Size	64 × 64	128 × 128	256 × 256	512 × 512	1024 × 1024	2048 × 2048
$\|M_p\|$	174	398	774	1630	3310	5712
$\|GD\|$	26	45	73	114	186	261
ER	6.9832	2.3311	1.2191	0.3763	0.141	0.0696
θ	181.5632	104.8995	88.9943	42.8982	26.226	18.1656

$|M_p|$, $|GD|$ denote the number of vertices of polygon M_p, the number of the geodesic diameter (GD) vertices, respectively, ER denotes the length error in % related to the length of γ obtained by the Gaussian quadrature, and $\theta = ER \times |GD|$ represents the measure of effectiveness of the polygonal approximation

Theorem 4 is a consequence of the general Theorem 2 and shows the convergence of geodesic diameter lengths to the length of γ. It has been shown [23] that M_{p+k}, $k = 0, 1, \ldots$ fulfil the assumption of c-nondegeneracy and they form a set of nested polygons so that they fulfil the condition of Theorem 2.

Let $\gamma\colon [a, b] \to R^2$ be a continuously differentiable planar Jordan arc which has the form

$$x(t) = 5\sin(t + 90) + 2\sin(3t + 50)$$

$$y(t) = 6\sin(t + 40 + 2\sin(2t + 70), \qquad t \in [0, \pi]. \tag{7}$$

The approximation of (7) by the geodesic diameter (GD) on different square grids shown in Table 2 in corresponding columns.

The resulting shortest polygonal Jordan curve which corresponds to the 512 × 512 square grid points resolution is shown in Fig. 9.

The number of extremal convex and concave vertices of polygonally bounded sets which correspond to planar Jordan curves obtained by above-mentioned gridding technique is limited by the number of extremal points of the curves and this

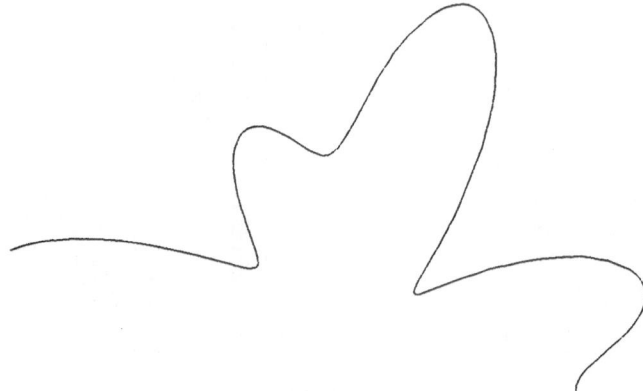

Figure 9

number is independent from the grid size density. The above-mentioned topological approach for planar Jordan curves and arcs approximation has a direct impact also on line drawing image processing.

5. Line Drawing Image Processing

In the previous section the problem of approximation of planar Jordan curves and planar Jordan arcs has been considered. This approximation is based on basic notions of intrinsic geometry of metric spaces, on the notion of the minimum perimeter polygon and on the notion of the geodesic diameter in a polygon which is a direct generalization of the notion of a diameter of a convex polygon. In the case of a gridding technique both represent a polygonal path whose vertices are grid points, so that they represent a vectorization process on a given grid which is feasible, i.e., the solution lies inside a polygonally bounded compact set or inside a polygon. Nevertheless, they represent the smoothest feasible polygonal paths, so that they represent efficient visualization techniques.

The trace of a planar Jordan curve or a planar Jordan arc is infinitesimally thin. To each such a curve corresponds on a given orthogonal grid so called bit map defined by the union of all squares which are crossed or touched by the given curve. This bit map represents a polygonally bounded compact set inside of which the given curve and also its polygonal approximation lie. The ultimate goal of a line drawing image processing system is to convert line drawings produced by an electronic drawing system into a highly compressed graphical representation. The drawings are represented by thin lines of fairly uniform thickness. The thickness could be arbitrary thin but they are not infinitesimally thin. To each line drawing corresponds again a bit map. This bit map is defined by the union of all squares of a given gray level intensity obtained by a suitable thresholding of the digital image obtained by capturing a line drawing image by a scanner. In the case of the reproduction of line drawing images it is assumed that plotted lines will be of the same thickness as original ones captured by the scanner.

The vectorization of a simple closed drawing line is performed by the minimum perimeter polygon which corresponds to the given bit map. The vectorization of a drawing line which does not touch or intersect itself (Jordan arc) is performed by the geodesic diameter in a polygon which corresponds to the given bit map. The vectorization of a line drawing image composed of closed and open drawing lines which do not touch or intersect each other consists then of the following steps:

a) thresholding
b) connectivity analysis
c) component labelling
d) x-axis parallel pseudodiagonalization
e) minimum perimeter polygon and/or geodesic diameter calculation.

A line drawing image is captured by a scanner and the resulting gray level image is thresholded by a suitable threshold level into a binary one [19, 32] which represents a bit map. An $N \times N$ binary image I_p defined on the square grid (5) is represented by

$$I_p = \bigcup_1^k N^p_{(w_1, w_2)}, k = N \times N,$$

where the color of $N^p_{(w_1, w_2)}$ which belong to the background is white and black $N^p_{(w_1, w_2)}$ squares belong to connected sets to be investigated. Each square $N^p_{(w_1, w_2)}$ is represented by four vertices (i, j), $(i + 1, j)$, $(i + 1, j + 1)$, $(i, j + 1)$ coordinates of which are integers. A row linear segment is the union of all edge connected squares on an x-axis parallel grid line and it represents a rectangle. Two row linear segments (i, j), (m, j), $(m, j + 1)$, $(i, j + 1)$ and $(k, j - 1)$, $(l, j - 1)$, (l, j), (k, j) with $l > k$, $m > i$ are edge connected if $k < m \wedge l > i$, see Fig. 10.

Connectivity analysis [6, 18, 20] is the most important procedure on binary images. It is a sequential procedure performed on an image from top to bottom.

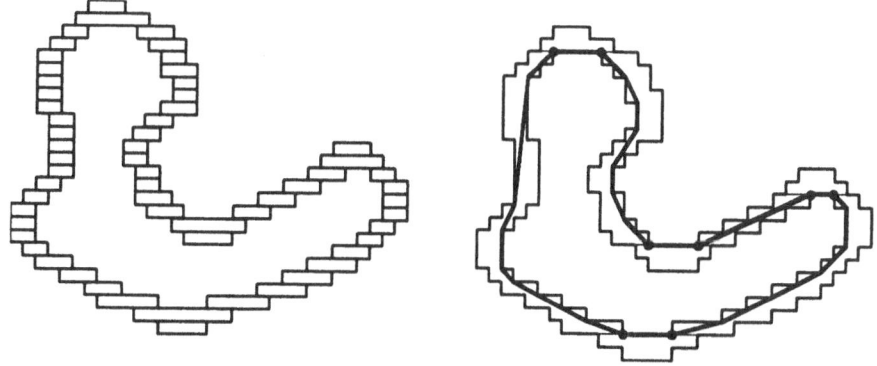

Figure 10

The analysis performs comparison of linear segments on a grid line with linear segments on the next grid line. It performs component labelling and it allows to count the number of connected components. In the case of line drawing image processing we will consider connected components which represent polygons or polygonally bounded compact sets with one hole. The x-axis parallel pseudodiagonals of a connected component which represents a polygonally bounded compact set can be found during the connectivity analysis procedure, see Fig. 10. It means that in the case of line drawing image processing for geodesic diameter and for the shortest path calculation it is not necessary to perform trapezoidization of a polygon or a polygonally bounded compact set. According to Algorithm (2) when x-axis parallel pseudodiagonals and extremal vertices have been found it is necessary to eliminate monotone polygons whose boundaries contain only vertices of L_1 or L_2, respectively, and to find the shortest path in each monotone polygon between two extremal vertices of resulting polygonally bounded set, see Fig. 10.

On Fig. 11 it is shown a city map and Fig. 12 shows its vectorized form, Fig. 13 shows an original map and Fig. 14 shows its vectorized form. The vectorized forms are represented by minimum perimeter polygons and geodesic diameters corre-

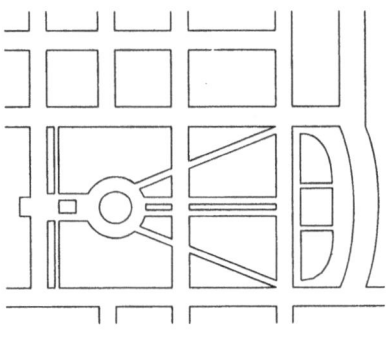

Figure 11

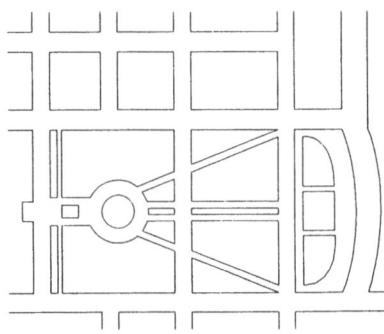

Figure 12

Figure 13

Figure 14

sponding to the bit maps of the line drawing images. The minimum perimeter polygon and a geodesic diameter, owing to their properties, represent the smoothest feasible vectorization of planar Jordan curves and planar Jordan arcs. The algorithms are based on integer arithmetic so that they are computationally fast. This vectorization represents a generalized chain code which was originally proposed for four or eight vectors [9].

6. Conclusion

Piecewise linear approximation of planar Jordan curves and arcs is described. This approximation represents a topological approach which allows to use by efficient ways so called gridding techniques. The algorithms for the planar Jordan curves and arcs approximation are based on the basic notions of intrinsic geometry of metric spaces: on the notion of the shortest path in a polygonally bounded compact set and on the notion of geodesic diameter of a polygon. Applications of this approach for line drawing image processing are highlighted.

Acknowledgement

The first author is indebted to the Alexander von Humboldt Stiftung for the financial support during his stay at the Institut für Angewandte Mathematik und Statistik in Würzburg.

References

[1] Alexandrov, A. D.: Intrinsic geometry of convex surfaces. Moscow: OGIZ, 1918 [in Russian].

[2] Allgower, E. L., Georg, K.: Numerical continuation methods. Berlin, Heidelberg, New York, Tokyo: Springer 1990.

[3] Chazelle, B., Incerpi, J.: Triangulation and shape-complexity. ACM Trans. Graph. 3, 135–152 (1984).

[4] Chazelle, B.: A theorem on polygon cutting with applications. Proc. 23-rd Annual Symp. Found. Comput. Sci. 339–349 (1982).

[5] Chazelle, B.: Triangulating a simple polygon in linear time. Discrete Comput. Geom. 6, 485–524 (1991).

[6] Dillencourt, M. B., Samet, H., Tamminen, M.: Connected component labeling for arbitrary binary representation. Proc. of the 5th International Conference on Image Analysis and Processing, pp. 131–146. Singapore: World Scientific 1990.

[7] Ferianc, P., Sloboda, F., Zat'ko, B.: On properties of the shortest path problem solution in a polygonally bounded compact set. Techn. Report of the Inst. of Control Theory and Robotics, Bratislava (1993).

[8] Fournier, A., Montuno, D. Y.: Triangulating simple polygons and equivalent problems. ACM Trans. Graph. 3, 153–174 (1984).

[9] Freeman, H.: On the encoding of arbitrary geometric configurations. IRE Trans. Electron. Comput. EC-10, 260–268 (1961).

[10] Garey, M. R., Johnson, D. S., Preparata, F. P., Tarjan, R. E.: Triangulating a simple polygon. Inf. Proc. Letters 7, 175–179 (1978).

[11] Hertel, S., Mehlhorn, K.: Fast triangulation of simple polygons. Proc. Conf. Found. Comput. Theory, pp. 207–218. New York: Springer 1983.

[12] Lee, D. T., Preparata, F. P.: Location of a point in a planar subdivision and its applications. SIAM J. Comput. 6, 594–606 (1977).

[13] Lee, D. T., Preparata, F. P.: Euclidean shortest paths in the presence of rectilinear barriers. Networks 14, 393–410 (1984).

[14] Lennes, N. J.: Theorems on the simple finite polygon and polyhedron. Am. J. Math. 33, 37–62 (1911).

[15] Montanari, U.: On limit properties of digitization schemes. J. ACM 17, 348–360 (1970).

[16] O'Rourke, J.: Art gallery theorems and algorithms. New York, Oxford: Oxford University Press 1987.

[17] Rheinboldt, W. C.: Solution fields of nonlinear equations and continuation methods. SIAM J. Numer. Anal. 17, 221–237 (1980).

[18] Rosenfeld, A., Pfaltz, J. L.: Sequential operations in digital image processing. J. ACM 13, 471–494 (1966).

[19] Rosin, P. L., West, G. A. W.: Segmentation of edges into lines and arc. Image Vision Comput. 7, 109–114 (1989).

[20] Sameth, H.: Connected component labeling using quadtrees. J. ACM 28, 487–501 (1981).

[21] Sklansky, J., Chazin, R. L., Hansen, B. J.: Minimum perimeter polygon of digitized silhouettes. IEEE Trans. Comput. 21, 260–268 (1972).

[22] Sloboda, F., Stoer, J.: On piecewise linear approximation of planar Jordan curves. J. Comput. Appl. Math. 55, 369–383 (1994).

[23] Sloboda, F., Stoer, J.: On piecewise linear approximation of planar Jordan arcs. Techn. Report of the Inst. of Control Theory and Robotics, Bratislava 1994.

[24] Sloboda, F., Zat'ko, B.: On linear time algorithm for the shortest path problem in a polygonally bounded compact set. Techn. Report of the Inst. of Control Theory and Robotics, Bratislava 1993.

[25] Sloboda, F., Zat'ko, B.: On properties of the shortest path problem solution in a polygon. Techn. Report of the Inst. of Control Theory and Robotics, Bratislava 1994.

[26] Sloboda, F., Zat'ko, B.: On shortest path problem solution in a polygon. Techn. Report of the Inst. of Control Theory and Robotics, Bratislava 1994.

[27] Suri, S.: The all-geodesic-furthest neighbors problem for simple polygons. Proc. Third Annual ACM Symposium on Computational Geometry, 64–75 (1987).
[28] Tarjan, R. E., Van Wyk, Ch. J.: An $O(n \log \log n)$-time algorithm for triangulating a simple polygon. SIAM J. Comput. *17*, 143–177 (1988).
[29] Toussaint, G. T.: An optimal algorithm for computing the relative convex Hull of points in a polygon. In: Signal processing III: theories and applications (Young, I. T., et. al., eds.), pp. 853–856. Amsterdam: North-Holland 1986.
[30] Toussaint, G. T.: Computing geodesic properties inside a simple polygon. Rev. Intell. Art. *3*, 9–42 (1989).
[31] Toussaint, G. T.: Efficient triangulation of simple polygons. Visual Comput. *7*, 280–295 (1991).
[32] Watson, L. T., Arvind, K., Ehrich, R. W.: Extraction of lines and regions from grey tone line drawing images. Pattern Rec. *17*, 493–507 (1984).

Dr. F. Sloboda
Dr. B. Zat'ko
Institute of Control Theory and Robotics
Slovak Academy of Sciences
Dúbravská 9
84237 Bratislava
Slovakia
e-mail: utrrbrit@savba.sk

Computing Suppl. 11, 201–220 (1996)

Segmentation with Volumetric Part Models*

F. Solina, Ljubljana

Abstract

Segmentation with Volumetric Part Models. Volumetric models are top-level shape representation in computer vision applications. Volumetric models are especially suited for part-level representation on which manipulation, recognition and other reasoning can be based. The two most popular types of volumetric models in computer vision are generalized cylinders and superquadrics. This paper gives an overview of recovery and segmentation methods applying these two types of volumetric models. Methods of segmentation into parts are analyzed and advantageous properties of part-models discussed.

Key words: Shape representation, superquadrics, generalized cylinders, part recovery.

1. Introduction

The goal of computer vision is to enable intelligent interaction of artificial agents with their surroundings. The means of this interaction are images of various kinds; intensity images, pairs of stereo images, range images or even sonar data. Images which at the sensory level consist of several hundreds or thousands of individual image elements must in this process be encoded in a more compact fashion. For any reasoning or acting on the surroundings, it is advantageous that this coding of images as well as the internal representation of the observed scene closely reflects the actual structure. Distinct objects, for example, should have distinct models of themselves. In this way, the labeling of individual entities, necessary for control and higher level reasoning, becomes possible.

So far, many different models have been used for modeling different aspects of objects and scenes. Models for representing 3D structures can be grouped into local and global models. Methods for local representation attempt to represent objects as sets of primitives such as surface patches or edges. Global methods on the other hand attempt to represent an object as an entity in its own coordinate system. When objects of such global models correspond to perceptual equivalents

* This research was supported in part by The Ministry for Science and Technology of The Republic of Slovenia (Projects P2-1122 and J2-6187).

of parts, we speak of part-level models. Several part-level models are required to represent an articulated object. A part-level shape description supports spatial reasoning, object manipulation, and structural object recognition. People often resort to such part description when asked to describe natural or man-made objects [37]. Such part descriptions are generally suitable for path planning or manipulation—for object-recognition, however, they are sometimes not malleable enough to represent all necessary details and several researchers are looking into extending part-level models with additional layers of details.

To obtain part-level descriptions of a scene *two* tasks must be accomplished. The image must be partioned into areas corresponding to individual parts—a problem referred to as segmentation—and recovering a part model for each of those segments. Normally, these two tasks are separated so that segmentation is performed first and then followed by modeling of previously isolated segments. In this way, segmentation cannot take directly into account the shapes that part-models can adopt. Adequate part-models for all such image segments may not exist in the selected modeling language. To avoid this problem segmentation and part-model recovery can be combined so that images are segmented only into parts which are instantiations of selected part-models.

The two most popular types of volumetric models employed in computer vision are generalized cylinders and superquadrics. Several other global models exist, however, that attempt to represent an object as an entity in their own coordinate system: Spherical harmonic surfaces [43], Gaussian images and extended Gaussian images [26], Symmetry seeking models [49], Blobby model [35], and Hyperquadrics [23]. In this paper we concentrate on the two most commonly used models: generalized cylinders and superquadrics. We compare them with regard to recovery and segmentation. We show that superquadrics are in respect to recovery from images advantageous in comparison to generalized cylinders, especially when segmentation and model recovery are to be combined.

The rest of the paper is divided as follows: the second section is on segmentation and ways of combining segmentation with model recovery. The following two sections describe model recovery and segmentation techniques for generalized cylinders and superquadrics. The fifth section discusses some typical applications of volumetric models in computer vision. In the last section general properties of part-level models are discussed.

2. Segmentation

Basic scientific methodology instigates decomposition of complex objects into parts, units or primitives to enable its study at various levels of abstraction. Abstraction is a crucial mechanism to cope with limits on how much information one can process at a time. In a similar way, to comprehend images, they should be decomposed into "natural" and "simple" parts that order and partition visual information into a limited number of perceptually significant parts. This challenging problem is called segmentation. Essential to segmentation is that the resulting parts ought to correspond to the underlying physical part structure of the scene depicted in the image. This is a prerequisite for image understanding. Part-level description of an image is therefore a necessary step towards building the scene description in terms of symbolic entities.

Segmentation entails decomposing images into segments so that each piece of information in an image is mapped either to a segment or discarded as noise. To get as compact a description as possible a minimum number of such part primitives should be used. To define what is "natural" and "simple" is a hard problem. It depends on the type of the observed scene as well as on the objective of the observing agent. The general segmentation problem is very difficult since multiple sources of image information should be involved. In this article we restrict the problem to shape information alone. In absence of domain knowledge, ambiguities can arise due to multiple representations, incomplete data, and multiple degrees of freedom of part-models.

In general, a criterion for segmentation must be defined. Standard segmentation criteria are different measures of homogeneity or difference since the two basic approaches to segmentation are:

1. finding of homogeneous regions, and
2. finding of borders (differences) between regions.

Segmentation methods based on homogeneity criterion can be implemented using different techniques, from simple tresholding, region growing, split and merge to scale-space approaches. Border-based segmentation employs different edge detection methods such as gradient methods, Hough transform, and active contour models.

To reach the part-model level representation that we discussed in the introduction, tokens that were obtained with such segmentation methods must be evaluated, filtered, grouped, and combined. Perfect segmentation into part-models would result only in segments that can be described with selected part-models, be it generalized cylinders or superquadrics. A common problem, however, is that the above-mentioned low-level segmentation methods often give part boundaries that cannot be modeled adequately with selected part-models. The cause of this problem is that grouping and combining of the low-level image elements do not evaluate the results on the basis of the final part-level model but on some lower level model or criterion (lines, contours, cross-sections, symmetries, aspect ratios, compactness etc.).

To overcome this problem, Bajcsy et al. [3] argued that segmentation and part-level modeling should be combined. A general method for combining segmentation and shape-recovery, called "recover-and-select" was developed by Aleš Leonardis [28]. In this article we show that when segmentation is combined with part-model recovery, low-level segmentation can be skipped over and fitness to given part-level shape models used directly as a segmentation criterion.

Hence, we divide segmentation methods into two general categories:

1. *Segment-then-Fit* methods, and
2. *Segment-and-Fit* methods.

Most standard segmentation methods belong to the first category which separates segmentation and shape recovery. This separation accounts for the above discussed problem. The methods in the second category use the final shape models also as a segmentation criterion and in this way achieve better segmentation results.

As mentioned in the Introduction we concentrate in this article on two types of part-level models: generalized cylinders and superquadrics. In the following two sections, which describe segmentation and part-modeling approaches using generalized cylinders and superquadrics, we will try to apply the above defined classification of segmentation methods.

3. Generalized Cylinders

The first dedicated volumetric models in computer vision were generalized cylinders. Generalized cylinders, sometimes referred to also as generalized cones, were defined by Thomas O. Binford in 1971 [8]. Generalized cylinder representation was preceded by earlier concepts for volumetric representation, notably the symmetric axis representation, developed by Blum [9]. Symmetric axis descriptions were defined as the set of centers of maximal spheres contained within the shape. All these representations are especially good at representing naturally evolved or grown elongated shapes.

A generalized cylinder is expressed by a volume obtained by sweeping a two-dimensional set or volume along an arbitrary space curve (Fig. 1). The set may vary parametrically along the curve. Different parameterizations of the above definition are possible. In general, a definition of the axis and the sweeping set are required. The axis can be represented as a function of arc length s in a fixed coordinate

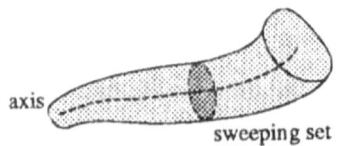

axis

sweeping set

Figure 1. A generalized cylinder is constructed by sweeping a closed contour along a space curve

system x, y, z

$$a(s) = (x(s), y(s), z(s)).$$

The sweeping set is more conveniently defined in a local coordinate system, defined at the origin of each point of the axis $a(s)$. The sweeping set can be defined by a cross section boundary, parameterized by another parameter r:

$$\text{sweeping set} = (x(r, s), y(r, s)).$$

This general definition is very powerful and a large variety of shapes can be described with it. In the most general form the generalized cylinder representation is so powerful that almost arbitrarily formed complete objects can be modeled. But since their outset generalized cylinders were used mainly as part-models [33], especially if restrictions to the general definition were applied (Fig. 2). To limit the complexity and simplify the recovery of generalized cylinder models from images researchers often restricted generalized cylinders to straight axes with constant sweeping sets. Properties of straight homogeneous general cylinders are addressed in [41].

Various methods for deriving descriptions based on generalized cylinders mostly from range images were attempted. One of the most widely publicized vision systems based on generalized cylinders was the ACRONYM system by Brooks [13]. The ACRONYM system was capable of recognizing different types of airplanes from intensity images of airports taken from the air.

Generalized cylinders influenced much of the model-based vision research in the past two decades. Besides building actual vision systems, generalized cylinders had some impact also on vision theory. GEONS introduced by Biederman [7] were conceived as a limited set of different primitive building blocks that could build any natural or man-made shape. The concept of geons was derived from qualitative changes of generalized cylinders. Geons are classified only on the basis of axis shape, cross-section shape, cross-section sweeping function, and cross-section symmetry. These qualitative geometrical properties could prove to be very useful in indexing object databases.

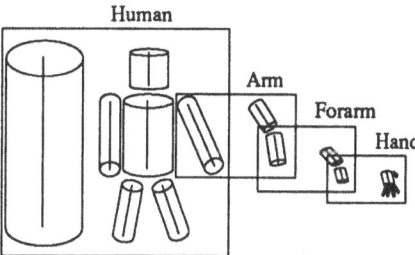

Figure 2. A hierarchical part-level representation of a human form with generalized cylinders as part-models (after [36])

3.1. Recovery of Generalized Cylinders

Recovery of generalized cylinders was undertaken from range images, pairs of stereo images, and from contours derived from single intensity images. Especially recovery from contours received the widest interest because humans are so good at perceiving a shape from its boundary.

Recovery of generalized cylinders has been approached from two broad perspectives:

1. In the first approach, some property of the 3-D surface is associated with each interpretation. The interpretation associated with the highest value of the selected property is chosen. Different properties were proposed as the preference criterion: *smoothness of the curve* [6], *curvature* [53], and *compactness* [12]. The smoothness property is a combination of the curvature and torsion of a 3-D curve. Over all possible 3-D curves that can generate a given 2-D image curve, the smoothest one is selected, or alternatively, the most compact 2-D shape. These methods in general require smooth and complete boundaries, some even implicitly assume planar contours.
2. The second approach to recovery of generalized cylinders are constraint-based. Constraints on 3-D surface orientations are inferred from a variety of observations (skew and parallel symmetry [27, 51]) and their propagation to other parts of the image (tilt and slant [48]). A unique solution is expected when various such constraints are combined.

Most recovery methods are restricted to straight homogeneous general cylinders (SHGC's) [51]. In general, one can criticize the methods of recovering generalized cylinders on the count that they are not very robust because they must rely on complicated rules for grouping of low level image models (i.e. edges, corners, surface normals) into models of larger granularity (i.e. symmetrical contours or cross-sections) to arrive finally to generalized cylinders. The recovery methods are sensitive to noise and often require complete data without any occlusions. These problems are due in part to the complicated parameterization of generalized cylinders and to the lack of a fitting function that would enable a straightforward numerical examination of the model's appropriateness for the modeled image data.

3.2. Segmentation with Generalized Cylinders

Most systems that recover generalized cylinders separate segmentation and model recovery and can be classified as "Segment-then-Fit" methods. The ACRONYM system, for example, finds first "ribbons", which are two-dimensional specializations of generalized cylinders, by an edge linking algorithm. By a system of forward and backward constraints these ribbons can be later matched to generalized cylinders which make up the model base.

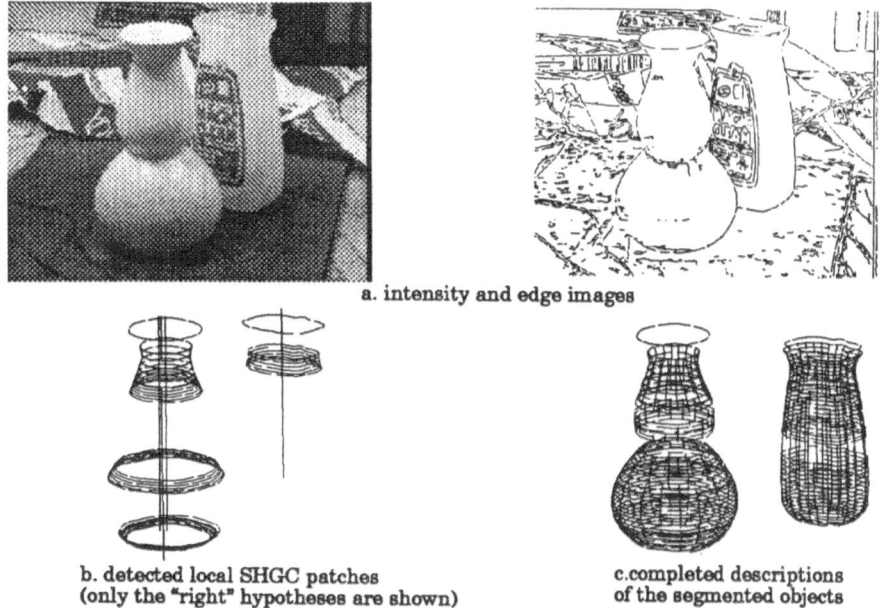

a. intensity and edge images

b. detected local SHGC patches
(only the "right" hypotheses are shown)

c.completed descriptions
of the segmented objects

Figure 3. Combined segmentation and recovery of straight homogeneous generalized cylinders (from [56])

Since generalized cylinders do not have a corresponding implicit function, a direct numerical evaluation of the model's fitness to the data is very demanding. This is why an integrated approach to shape recovery and segmentation is difficult. Notwithstanding these obstacles, the latest results in recovery and segmentation of straight homogeneous generalized cylinders (SHGC) achieved by Zerrough and Nevatia are encouraging [56] (Fig. 3). The method which integrates segmentation and recovery of SGHCs is quite complex and requires three grouping levels: the curve level, the parallel symmetry level, and the SHGC patch level. The SHGC level is intended to form complete SHGC object descriptions whenever possible in the image. Constraints used in all three grouping levels are derived from geometric projective properties of SHGCs.

4. Superquadrics

Superquadric models appeared in computer vision as an answer to some of the problems with recovery of generalized cylinders [37]. Superquadrics are solid models that can, with a fairly simple parameterization, represent a large variety of standard geometrical solids as well as smooth shapes in between. This makes them convenient for representing rounded, blob-like shaped parts, typical for objects formed by natural processes.

To introduce the concept of superquadrics we define their 2-D equivalent. A superellipse is a closed curve defined by the following simple equation:

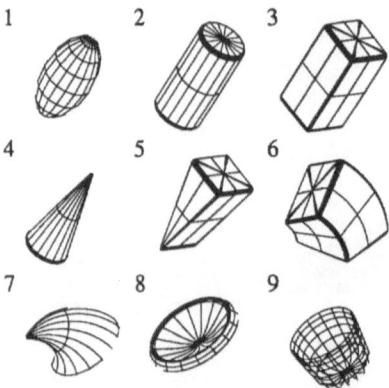

Figure 4. Superquadric models enhanced with global deformations (from [47])

$$\left(\frac{x}{a}\right)^m + \left(\frac{y}{b}\right)^m = 1,$$

where a and b are the size (positive real number) of the major and minor axes and m is a rational number

$$m = \frac{p}{q} > 0, \quad \text{where} \begin{cases} p & \text{is an even integer,} \\ q & \text{is an odd integer.} \end{cases}$$

If $m = 2$ and $a = b$, we get the equation of a circle. For larger m, however, we get gradually more rectangular shapes, until for $m \to \infty$, the curve takes up the shape of a square. Superellipses are special cases of curves which are known in analytical geometry as Lamé curves [31][1]. Piet Hein, who popularized these curves for design purposes, also made a generalization to 3D which he named *superellipsoids* or *super-spheres* [17]. The final mathematical foundation of superquadrics was laid out by Barr [4], who generalized the whole family of quadric surfaces with the help of varying exponents, and coined a new name for them—*superquadrics*. Super-quadrics are by definition a family of shapes that includes not only superellipsoids, but also superhyperboloids of one and of two pieces, as well as supertoroids (see Fig. 4).

The explicit superellipsoid equation, defined by the following surface vector, is

$$\mathbf{x}(\eta, \omega) = \begin{bmatrix} a_1 \cos^{\varepsilon_1}(\eta) \cos^{\varepsilon_2}(\omega) \\ a_2 \cos^{\varepsilon_1}(\eta) \sin^{\varepsilon_2}(\omega) \\ a_3 \sin^{\varepsilon_1}(\eta) \end{bmatrix} \quad \begin{array}{l} -\pi/2 \le \eta \le \pi/2 \\ -\pi \le \omega < \pi \end{array}, \tag{1}$$

---

[1] Lamé curves are named after the French mathematician Gabriel Lamé, who was the first who studied them in *Examen des différentes méthodes employées pour résoudre les problémes de geometrie*, Paris, 1818.

where a_1, a_2 and a_3 determine size, and ε_1 and ε_2 determine global shape. The alternative, implicit superellipsoid definition, also called the *inside-outside* function is

$$\left(\left(\frac{x}{a_1} \right)^{2/\varepsilon_2} + \left(\frac{y}{a_2} \right)^{2/\varepsilon_2} \right)^{\varepsilon_2/\varepsilon_1} + \left(\frac{z}{a_3} \right)^{2/\varepsilon_1} = 1. \tag{2}$$

Points x, y, z that correspond to the above equation are on the surface of the superellipsoid.

For numerical calculation, it is easier to assume that exponents ε_1 and ε_2 can be any positive real number and not only rational numbers with an even enumerator. Then, one should assume that exponentiation in Eqs. (1) and (2) means

$$x^p = \text{sign}(x)|x|^p = \begin{cases} x^p & x \geq 0 \\ -|x|^p & x < 0 \end{cases}$$

to avoid complex numbers when a negative real number is raised to a real exponent. For applications in computer vision, the values for ε_1 and ε_2 are normally bounded: $2 > \varepsilon_1, \varepsilon_2 > 0$, so that only convex shapes are produced. For a superquadric in canonical position one needs to set the value of 5 parameters (3 for size in each dimension, 2 for shape defining exponents). For a superquadric in general position 6 additional parameters are required to define the translation and rotation of the model.

Superquadric models, which compactly represent a continuum of useful forms with rounded edges, and which can easily be rendered and shaded due to their dual normal equations, and deformed by parametric deformations, are very useful in computer graphics. Parametric deformations such as twisting, bending, tapering, and their combinations can enhance the expressive power of superquadrics [5]. Parametric deformations typically require just a few more parameters.

Pentland [37] was the first who grasped the potential of the superquadric models and parametric deformations for modeling natural shapes in the context of computer vision. He proposed to use superquadric models in combination with global deformations as a set of primitives which can be molded like lumps of clay to describe the scene structure at a scale that is similar to our naive perceptual notion of *parts*. Pentland presents several perceptual and cognitive arguments to recover the scene structure at such a part-level since people seem to make heavy use of this part structure in their perceptual interpretation of scenes. The superquadrics, which are like phonemes in this description language, are deformed by stretching, bending, tapering or twisting, and then combined, using Boolean operations to build complex objects. In general, the same arguments as the ones for using generalized cylinders as part models hold for superquadrics.

Superquadrics are in fact a subset of generalized cylinders. Any superquadric can be in principle represented as a generalized cylinder admitting that the parameterization could be much more complicated. The geon concept can also be expressed in terms of superquadrics. Raja and Jain [42] conducted experiments on mapping superquadric shapes to 12 shape classes corresponding to a "collapsed" set of 36 different geons. Raja and Jain used the five shape and deformation parameters of superquadrics for classification into 12 geon classes using binary tree and k-nearest-neighbor classifiers.

4.1. Recovery of Superquadrics

The problem of recovering superquadrics from images is an overconstrained problem. A few model parameters (i.e. 11 for non-deformed superquadrics) must be determined from several (i.e. a few hundred) image features (range points, surface normals or points on occluding contours). By its parameterization the superquadrics impose a certain symmetry and in this way place some reasonable constraints on the shape of the occluded portion of a three dimensional object.

In the first article on the use of superquadrics in computer vision, Pentland [37] proposed an analytical method for recovery of superquadrics using the explicit equation (1). Except for some simple synthetic images, this analytical approach did not turn out to be feasible. Pentland [38] later proposed another method which combined recovery with segmentation and was based on a coarse search through the entire superquadric parameter space for a large number of overlapping image regions. The major objection to this method is its excessive computational cost.

Iterative methods based on non-linear least squares fitting techniques using different distance metrics were proposed [2, 10]. Solina and Bajcsy [2, 47] formulated the recovery of deformed superquadric models from pre-segmented range data as a least-squares minimization of a fitting function. An iterative gradient descent method was used to solve the non-linear minimization problem. A modified superquadric implicit or inside-outside function (Eq. (2)) with an additional multiplicative volume factor was used as the fitting function. The volume factor is used to ensure the recovery of the *smallest* superquadric model that fits the range data in the least squares sense. To the standard superquadric model, which requires 11 parameters, linear tapering, bending, and a cavity deformation were added, which adds up to a total of 18 parameters. Recovery of a single superquadric model requires on the average about 30 iterations (see Fig. 5).

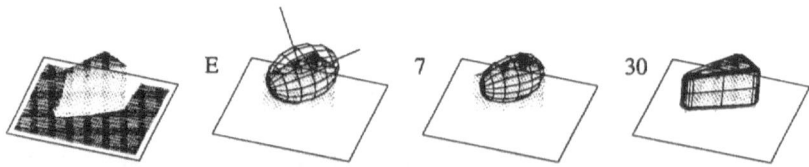

Figure 5. Recovery of a tapered superquadric from pre-segmented range data (from [47]). From left to right: original range image; *E* initial estimate; *7, 30* models after the 7th and 30th iteration

Pentland [39] proposed another superquadric recovery method. Segmentation was first achieved by matching 2D silhouettes (2D projections of 3D superquadric parts of different shapes and of different orientations) to the image data. After part segmentation, superquadric models were fitted to range data of individual part regions. Superquadric fitting based on modal dynamics [40] used as the error metric the squared distance along the depth axis z between the range data and the projected volume's visible surface.

Hager [22] combined the estimation (recovery) process of superquadric models with the decision-making process (i.e. graspability, categorization). Usually both processes are divorced in the sense that first a recovery process is performed and then a decision is made based on the recovered models. Combining both stages should result in minimal work required to reach a decision. The approach is based on a interval-bisection method to incorporate sensor based decision making in the presence of parametric constraints. The constraints describe a model for sensor data (i.e. superquadric) and the criteria for correct decisions about the data (i.e. categorization—see also [46]). An incremental constraint solving technique performs the minimal model recovery required to reach a decision. The major drawback of the method is slow convergence when categorization is involved. Determining the shape parameters ε_1 and ε_2 required several hundred iterations.

Yokoya et al. [57] experimented with simulated annealing to minimize a new error-of-fit measure for recovery of superquadrics from pre-segmented range data. The measure is a linear combination of distance of range points to the superquadric surface and difference in surface normals (first proposed by Bajcsy and Solina in [2]). Several hundred iterations were needed to recover models from range data.

Vidmar and Solina [52] studied the recovery of superquadrics from 2D contours. For a given contour several possible superquadric interpretations were derived. To a human observer some of these interpretations are obviously more natural than others, although all recovered models have a very tight fit to the contour data. Perceptually better solutions could be selected by using just a few additional pieces of information (a few range points or shading information).

Horikoshi and Suzuki [25] multiplied the objective function with a weighting function for robust estimation (based on whether the point is closer to the median value of the inside-outside function, or far from it in either directions). Consequently, the model is less sensitive to outliers.

Since no direct comparison of different metrics and minimization methods for superquadric recovery was made, it is difficult to rank the recovery methods only on the basis of results presented in the articles. Some experimental comparisons of different error-of-fit measures are given in [18]. Gupta [20] also discusses the error-of-fit functions. What is finally important is the perceptual likeness of models to the actual objects, the speed of convergence, and last but not least, the simplicity of implementation. On this ground the method proposed in [47] received a wide acceptance since several other authors have used it in their vision or robotic systems [1, 16, 19, 20, 30, 42].

4.2. Segmentation with Superquadrics

The early work on superquadrics concentrated primarily on recovery of isolated parts and did not address segmentation or used superquadrics as volumetric primitives *after* segmentation was achieved. Once recovery of superquadrics was well understood, more sophisticated techniques were designed that applied superquadrics to scene segmentation [19, 25, 30].

4.2.1. Two-Stage Segment-then-Fit Methods

These two stage methods decouple segmentation and model recovery. First, images are segmented, then part-models are fitted to the resulting regions.

Pentland [39], for example, used matched filters to segment binarized image data into part regions. The best set of binary patterns that would completely describe a silhouette is selected. The 3D data corresponding to each of the selected patterns was then fitted with a deformable superquadric based on modal dynamics.

Gupta et al. [21] used an edge based region growing method to segment range images of compact objects in a pile. The regions were segmented at jump boundaries, and each recovered region was considered a superquadric object. Reasoning was done about the physical support of these regions, and several possible 3D interpretations were made based on various scenarios of the object's physical support. A superquadric model was fitted and classified corresponding to each recovered object.

Ferrie et al. [16] used differential geometric properties and projected space curves modeled as snakes for segmenting range data. An augmented Darboux frame is computed at each point by fitting a parabolic quadric surface, which is iteratively refined by a curvature consistency algorithm.

Another qualitative shape recovery method using geon theory was proposed by Metaxas and Dickinson [34] to recover superquadrics on intensity data. Their integrated method uses Dickinson et al.'s geon-based segmentation scheme [14] (into ten geon classes) to provide orientation constraint and edge segments of a part. This is then the input to the physically-based global superquadric model recovery scheme developed by Terzopoulos and Metaxas [50].

All two-stage approaches suffer from the problem that the results of segmentation might not correspond tightly to any superquadric model. Thus, model recovery on such part domain will be uncertain about the shape, size and orientation of the model. To describe adequately a scene with a particular shape language, one should use this language also for partitioning the scene. In this case, a method that can accommodate the segments to the orientation, shape, and size of the superquadric model must be used. This can be accomplished by interleaving model recovery and segmentation [3].

4.2.2. Interleaved Segment-and-Fit Methods

Solina first attempted part-segmentation with superquadrics by recursively splitting the modeling domains and rejecting extraneous range points during model recovery [44]. It was, however, extremely difficult to constrain the single model recovery to take part-structure into account. Clearly, segmentation into part models must recover parts by hypothesizing parts and testing (evaluating) them.

Pentland [38] was the first one to successfully integrate segmentation and superquadric model recovery. However, the brute-force method of searching the entire parameter space of superquadrics for a large number of overlapping image regions is computationally excessively expensive.

Gupta and Bajcsy [19], proposed a recursive global-to-local technique for part recovery. A global model is recovered for the entire data (using [47]), which is evaluated by studying local and global residuals so that the further course of action can be determined. A set of qualitative acceptance criteria define suitability of the model. If the model is found to be deficient in representing data, then additional part models are hypothesized on the un-described data (regions of surface underestimation). The global model is refitted on the remaining data. Thus, the global model shrinks while the part models grow, yielding a hierarchical part-structure.

A tighter integration of segmentation and model recovery was achieved by combining the "recover-and-select" paradigm developed by Leonardis [28, 29] with the superquadric recovery method by Solina [44]. This work demonstrates that superquadrics can be *directly* and *simultaneously* recovered from range data without the mediation of any other geometric models [30]. The *recover-and-select* paradigm for the recovery of geometric parametric structures from image data [28] was originally developed for the recovery of parametric surfaces [29]. The paradigm works by recovering independently superquadric part models everywhere on the image, and selecting a subset which gives a compact description of the underlying data (Fig. 6).

Horikoshi and Suzuki [25] proposed a segment-and-merge method to segment 2D contours (with the figure of interest separated from the background) and sparse 3D data. This recursive procedure results in a possibly overlapping convex superquadric parts. Parts are then merged to arrive at a compact description.

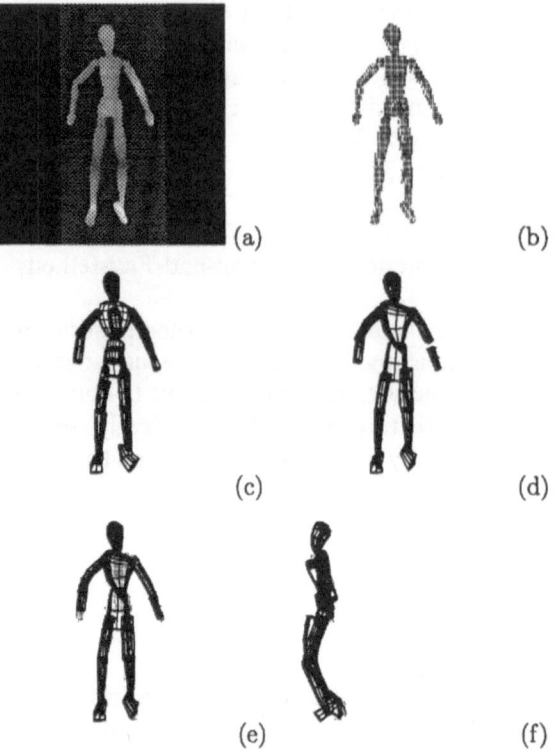

Figure 6. Range image segmentation using deformed superquadric part models. The human form was segmented by employing the *recover-and-select* paradigm of Leonardis [28] and the superquadric recovery method of Solina [44]. **a** Original range image, **b** input range points, **c, d** models during the recovery process, **e, f** the final result from two views

5. Application of Volumetric Models in Computer Vision

This section explores the role of volumetric shape primitives in computer vision tasks. Typical tasks that require volumetric shape primitives are object classification and object recognition. For simple object classification tasks which are required, for example, for grasping, sorting or object avoidance, part-models offer sufficient information.

The role of part-models in recognition depends on the type and number of objects that need to be recognized. Sometimes the structure and coarse shape of individual parts are enough. Often, the surface detail that the discussed part-models offer is poor for the required recognition task. As is the case with any shape primitive, superquadrics as well as straight homogeneous generalized cylinders have a limited shape vocabulary. They can be used to capture the global coarse shape of a 3D object or its constituent parts. The addition of global deformations increases the expressive power of superquadrics, but still limits it to the global coarse shape as opposed to local details. This lack of fine scale representation can be addressed by adding local degrees of freedom to superquadrics [40, 50], or relax the usual constraints on generalized cylinders. Terzopoulos et al. proposed a "symmetry-seeking model" for 3D object reconstruction which is a kind of a generalized cylinder that can accommodate even local "bumps" on the overall shape [49].

However, one drawback of such locally deformable extensions is that they have too many degrees of freedom to meaningfully segment even a simple scene. The increase in expressive power of part-models also results in an increase in complexity of all the visual tasks like segmentation, representation, recognition, and classification. Consequently, all of the segmentation and classification work with superquadrics [16, 19, 30, 39] has used globally deformable models, limiting the role of local deformations to refine the surface details.

The earliest works on superquadrics dealt primarily with single model analysis, since interpretation of a complex scene required model recovery to be understood first. These methods focussed either on classification of single models, where the power of superquadrics as a compact parametric model was exploited [24, 42, 46], or on using superquadrics as a volumetric primitive *after* a segmentation had been obtained [15, 21, 38]. Once the model recovery was understood, more sophisticated techniques were designed to apply superquadrics to scene segmentation [19, 25, 30].

Solina and Bajcsy [46] recovered objects in the postal domain and categorized them as flats, tubes, parcels, and irregular packages based on the shape and size parameters of the segmented recovered models. Gupta et al. [21] extended Solina's approach to work on a cluttered scene by segmenting the range image using an independent edge-based scheme, and then recovering individual postal objects after reasoning about the physical supporting plane to constrain the 3D shape of the object.

Horikoshi and Kasahara [24] partitioned the superquadric parametric space between $0 < \varepsilon_1 \leq 2.0$ and $1.0 \leq \varepsilon_2 \leq 3.0$ to develop a shape indexing language. They mapped the representation space to verbal instructions like "rounder", "pinch", "flatten", etc., and developed a man-machine interface to construct object models. They also described an indexing scheme where complex objects were stored as superquadric models and indexed by model parameters.

Model-driven recognition (with superquadric part-primitives) has not so far been exploited despite the compact representation. The reason is that it is very difficult to recover "canonical" representations of objects from real data. Instead of recovering canonical descriptions, most researchers have followed the data-driven bottom-up strategy of fitting superquadric models to the data. The recognition problem then reduces to matching the recovered superquadric parts with the superquadric parts in the model base.

6. Conclusions

Part-modeling is a convenient abstraction mechanism in image understanding and practiced in many different kinds of applications. This article focused in particular on two types of part-models: generalized cylinders and superquadrics. Methods for their recovery and their role in segmentation were reviewed.

Recovery of generalized cylinders and segmentation with generalized cylinders is hampered by complicated methods that require grouping and combination of various constraints which results in relatively slow implementations and this even with confined forms of generalized cylinders (SHGCs). Most of the research on recovery of generalized cylinders used contours as input.

Despite initial reluctance in using superquadrics due to their nonlinear form, they have proven to be the primitives of choice for many a researcher seeking a volumetric model. The interchangeable implicit and explicit superquadric equations enable numerical evaluation of the model directly on the range data and computation of other desirable shape properties (i.e. surface normals). Superquadric models have shown to be useful as volumetric shape primitives for object categorization, segmentation, recognition, and representation. "Segment-and-fit" method using superquadrics reports good results on objects which would otherwise be un-segmentable with surface-based techniques [30]. Interpretation of intensity and sparse 3D data with superquadrics is still an open problem.

For a fair comparison of generalized cylinders and superquadrics, straight homogeneous generalized cylinders and superquadrics enhanced with global deformations should be used since for these two types of part-models methods of recovery exist. To evaluate and compare different representations the following three criteria defined by Marr and Nishihara [33] can help:

1. *accessibility—can the desired description be computed from an image and can it be done economically?* Faster, conceptually more concise, and more reliable methods exist for recovery of superquadrics.
2. *scope and uniqueness—what class of images is the representation designed for?* Due to their capability of modeling rounded edges and corners superquadrics enhanced with global deformations offer a richer vocabulary for description of naturally occurring shapes than do SHGCs. On the other hand are superquadrics, unlike generalized cylinders, limited in the shape of the cross section along the main axis.
3. *stability and sensitivity—do differences between descriptions in the representation reflect the relative importance of differences between the images described with respect to the task at hand?* Generalized cylinders and superquadrics offer a relatively stable part-level decomposition. Although the shape of recovered part-models is normally stable, the local coordinate system is not necessarily stable. This means that the actual values of model parameters cannot be used directly for recognition. Sensitivity depends on the required level of detail and is coupled with the scale of the representation.

Using a parametric shape model for vision requires that model evaluation be built into the segmentation and recognition systems. To this end, it is imperative to study the residuals of shape models by comparing them against the given data. Whaite and Ferrie [55] have recently described a decision theoretic framework to evaluate the fitted models. They extended their earlier work [54] on uncertainty in model parameters to develop three lack-of-fit statistics. Gupta and Bajcsy [19] used global *and* local distribution of residuals to determine model fitness and segmentation options on static data.

Brady [11] proposed three additional criteria for evaluating representations:

1. *rich local support—representation should be information preserving and locally computable (local frames).* Generalized cylinders satisfy this since there is a natural local coordinate frame at all points along the spine [11]. Superquadrics are computable even on small patches of range data, called seeds in [30].
2. *frames—by smooth extension and subsumption local frames should give rise to more global descriptions called frames.* Smooth extension and subsumption is implicit in the definition of generalized cylinders [11]. How can this be exploited for recovery of generalized cylinders is unclear. A superquadric can extend (grow) in the process of segmentation until it reaches in the data the natural part limits which are allowed by its internal parameterization. Subsumption can be achieved by a selection mechanism based on the Minimal Description Criteria [30].
3. *propagation of frames—frames that correspond to perceptual sub parts of a shape can be propagated by inheritance or affixment.* In general, there are partially defined frames naturally associated with the general cylinders or superquadrics. The frame is partial in that certain choices are underconstrained or arbitrary. By inheritance or affixment constraints from adjoining parts can be propagated. In the segmentation method presented in [30] this is not implemented yet.

Real situations demand that the parametric shape models must be recovered on partial and noisy single viewpoint data. Data could be missing due to other objects occluding the view, or due to the shadows in scanner geometry, or due to self-occlusion in single viewpoint data [32]. Noise in 3D measurements is inevitable and most difficult to model. While the symmetry constraints of superquadrics are useful in predicting the missing information, the downside of parametric models is the lack of uniqueness in describing incomplete and noisy data within an acceptable error of tolerance. This fact is borne out in the experiments conducted by [54], where they derive an ellipsoid of confidence within which all the acceptable models lie. Therefore we propose two additional criteria to judge future shape representations:

1. *propagation of uncertainties—*the shape representation should support the representation and propagation of sensor and modeling uncertainties.
2. *use of perceptual preferences in case of insufficient data—*when due to occlusion or sensor characteristics the image data does not offer enough constraints, perceptual preferences should be utilized to restrain the model.

To summarize, because superquadrics have a nice and complete mathematical definition, more powerful mathematical tools are available for their recovery from images. Study of generalized cylinders is distinguished by theoretical rigor but lacks for now efficient methods for recovery from images.

References

[1] Allen, P. K., Michelman, P.: Acquisition and interpretation of 3-D sensor data from touch. IEEE Trans. Robotics Automation 6, 397–404 (1990).

[2] Bajcsy, R., Solina, F.: Three dimensional shape representation revisited. In: Proceedings of the First International Conference on Computer Vision, pp. 231–240. London: IEEE 1987.

[3] Bajcsy, R., Solina, F., Gupta, A.: Segmentation versus object representation—are they separable? In: Analysis and interpretation of range images (Jain, R., Jain, A., eds.), pp. 207–223. New York: Springer 1990.

[4] Barr, A. H.: Superquadrics and angle-preserving transformations. IEEE Comput. Graphics Appl. 1, 11–23 (1981).

[5] Barr, A. H.: Global and local deformations of solid primitives. Comput. Graphics 18, 21–30 (1984).

[6] Barrow, H. G., Tenenbaum, J. M.: Interpreting line drawings as three dimensional surfaces. Art. Intell. 17, 75–116 (1981).

[7] Biederman, I.: Human image understanding: recent research and theory. Comput. Vision Graphics Image Proc. 32, 29–73 (1985).

[8] Binford, T. O.: Visual perception by computer. In: Proceedings of the IEEE Conference on Systems and Control, Miami, FL, 1971.

[9] Blum, H.: Biological shape and visual science (Part I). J. Theor. Biol. 38, 205–287 (1973).

[10] Boult, T. E., Gross, A. D.: Recovery of superquadrics from depth information. In: Proceedings of Spatial Reasoning and Multi-Sensor fusion workshop, pp. 128–137. St. Charles: SPIE 1987.

[11] Brady, M.: Criteria for representations of shape. In: Human and Machine Vision (Beck, J., Hope, B., Rosenfeld, A., eds.), pp. 39–84. New York: Academic Press 1983.

[12] Brady, M., Yuille, A.: An extremum principle for shape from contour. IEEE Trans. Pattern Anal. Machine Intell. 6, 288–301 (1984).

[13] Brooks, R.: Model-based 3-D interpretation of 2-D images. IEEE Trans. Pattern Anal. Machine Intell. 5, 140–150 (1983).

[14] Dickinson, S. J., Pentland, A. P., Rosenfeld, A.: From volumes to views: An approach to 3-D object recognition. CVGIP: Image Understanding 55, 130–1154 (1992).

[15] Ferrie, F. P., Lagarde, J., Whaite, P.: Darboux frames, snakes, and superquadrics: Geometry from the bottom-up. In: Proc. IEEE Workshop on Interpretation of 3D Scenes, pp. 170–176, Austin, TX, 1989.

[16] Ferrie, F. P., Lagarde, J., Whaite, P.: Darboux frames, snakes, and superquadrics: Geometry from the bottom up. IEEE Trans. Pattern Anal. Machine Intell. 15, 771–784 (1993).

[17] Gardner, M.: The superellipse: a curve that lies between the ellipse and the rectangle. Sci. Amer. 213, 222–234 (1965).

[18] Gross, A. D., Boult, T. E.: Error of fit measures for recovering parametric solids. In: Proceedings of the 2nd International Conference on Computer Vision, pp. 690–694. Tampa: IEEE 1988.

[19] Gupta, A., Bajcsy, R.: Volumetric segmentation of range images of 3-D objects using superquadrics. CVGIP: Image Understanding, 58, 302–326 (1993).

[20] Gupta, A., Bogoni, L., Bajcsy, R.: Quantitative and qualitative measures for the evaluation of the superquadric models. In: Proc. IEEE Workshop on Interpretation of 3D Scenes, pp. 162–169. Austin, TX, 1989.

[21] Gupta, A., Funka-Lea, G., Wohn, K.: Segmentation, modeling and classification of the compact objects in a pile. In: Proceedings of the Conference on Intelligent Robots and Computer Vision VIII: Algorithms and Techniques, pp. 98–108, Philadelphia, PA, November 1989. St. Charles: SPIE.

[22] Hager, G. D.: Constraint solving methods and sensor-based decision making. In: Proceedings of the International Conference on Robotics and Automation, pp. 1662–1667, Nice, France, 1992. London: IEEE.

[23] Hanson, A. J.: Hyperquadrics: smoothly deformable shapes with complex polyhedral bounds. Comput. Vision Graphics Image Proc. 44, 191–210 (1988).

[24] Horikoshi, T., Kasahara, H.: 3-D shape indexing language. In: Proceedings of the International Conference on Computers and Communications, Scottsdale, AZ, 1990. London: IEEE.
[25] Horikoshi, T., Suzuki, S.: 3D parts decomposition from sparse range data using information criterion. In: Proceedings of the CVPR, pp. 168–173, 1993.
[26] Horn, B. K. P.: Robot vision. Cambridge/Mass.: MIT Press 1986.
[27] Kanade, T.: Recovery of the three-dimensional shape of an object from a single view. Art. Intell. *17*, 409–460 (1981).
[28] Leonardis, A.: Image analysis using parametric models: model-recovery and model-selection paradigm. PhD Diss., University of Ljubljana, Faculty of Electr. Eng. and Computer Science, 1993.
[29] Leonardis, A., Gupta, A., Bajcsy, R.: Segmentation as the search for the best description of the image in terms of primitives. In: Proc. Third Int. Conf. on Computer Vision, pp. 121–125, Osaka, Japan, 1990. London: IEEE.
[30] Leonardis, A., Solina, F., Macerl, A.: A direct recovery of superquadric models in range images using recover-and-select paradigm. In: Proc. Third European Conf. on Computer Vision, pp. 309–318, Stockholm, Sweden, 1994.
[31] Loria, G.: Spezielle algebraische und transzendente ebene Kurven. Leipzig, Berlin: B. G. Teubner 1910.
[32] Maver, J., Bajcsy, R.: Occlusions as a guide for planning the next view. IEEE Trans. Pattern Anal. Machine Intell. *15*, 417–433 (1993).
[33] Marr, D., Nishihara, K.: Objects, parts, and categories. Proc. R. Soc. London Ser. *B 200*, 269–294 (1978).
[34] Metaxas, D., Dickinson, S.: Integration of quantitative and qualitative techniques for deformable model fitting from orthographic, perspective, and stereo projections. In: Proceedings of International Conference on Computer Vision, pp. 641–649. London: IEEE 1993.
[35] Muraki, S.: Volumetric shape description of range data using "Blobby model". Comput. Graphics *25*, 227–235 (1991).
[36] Nishihara, H. K.: Intensity, Visible-Surface, and Volumetric Representations. Art. Intell. *17*, 265–284 (1981).
[37] Pentland, A. P.: Perceptual organization and the representation of natural form. Art. Intell. *28*, 293–331 (1986).
[38] Pentland, A. P.: Recognition by parts. In: Proceedings of the First International Conference on Computer Vision, pp. 612–620. London: IEEE 1987.
[39] Pentland, A. P.: Automatic extraction of deformable part models. Int. J. Comput. Vision *4*, 107–126 (1990).
[40] Pentland, A. P., Sclaroff, S.: Closed-form solutions for physically based shape modeling and recognition. IEEE Trans. Pattern Anal. Machine Intell. *13*, 715–729 (1991).
[41] Ponce, J., Chelberg, D., Mann, W. B.: Invariant properties of straight homogeneous generalized cylinders and their contours. IEEE Trans. Pattern Anal. Machine Intell. *11*, 951–966 (1989).
[42] Raja, N. S., Jain, A. K.: Recognizing geons from superquadrics fitted to range data. Image Vision Computing *10*, 179–190 (1992).
[43] Schudy, R. B., Ballard, D. H.: Towards an anatomical model of hearth motion as seen in 4-D cardiac ultrasound data. Proceedings 6th Conference on Computer Applications in Radiology and Computer-Aided Analysis of Radiological Images, 1979.
[44] Solina, F.: Shape recovery and segmentation with deformable part models. PhD. Diss., University of Pennsylvania, Department of Computer and Information Science, 1987.
[45] Solina, F., Bajcsy, R.: Range image interpretation of mail pieces with superquadrics. In: Proceedings AAAI-87, Vol. 2, pp. 733–737. Seattle: WA, 1987.
[46] Solina, F., Bajcsy, R.: Recovery of mail piece shape from range images using 3-D deformable models. Int. J. Res. Eng. (Inaugural Issue) 125–131 (1989).
[47] Solina, F., Bajcsy, R.: Recovery of parametric models from range images: The case for superquadrics with global deformations. IEEE Trans. Pattern Anal. Machine Intell. *12*, 131–147 (1990).
[48] Stevens, K. A.: The visual interpretations of surface contours. Art. Intell. *17*, 47–73 (1981).
[49] Terzopoulos, D., Witkin, A., Kass, M.: Symmetry-seeking models and 3D object recognition, Int. J. Comput. Vision *1*, 211–221 (1987).
[50] Terzopoulos, D., Metaxas, D.: Dynamic 3D models with local and global deformations: Deformable superquadrics. IEEE Trans. Pattern Anal. Machine Intell. *13*, 703–714 (1991).
[51] Ulupinar, F., Nevatia, R.: Perception of 3-D surfaces from 2-D contours. IEEE Trans. Pattern Anal. Machine Intell. *15*, 3–18 (1993).
[52] Vidmar, A., Solina, F.: Recovery of superquadric models from occluding contours. In: Theoretical

foundations of computer vision (Klette, R., Kropatsch, W., eds.), pp. 227–240. Berlin: Akademie Verlag, 1992 (Mathematical Research, Vol. 69).

[53] Weiss, I.: 3-D shape representation by contours. Comput. Vision Graphics Image Proc. *41*, 80–100 (1988).

[54] Whaite, P., Ferrie, F. P.: From uncertainty to visual exploration. IEEE Trans. Pattern Anal. Machine Intell. *13*, 1038–1049, October 1991.

[55] Whaite, P., Ferrie, F. P.: Active exploration: Knowing when we're wrong. In: Proceedings of the ICCV, pp. 41–48, Berlin, Germany, 1993. IEEE CS.

[56] Zerrough, M., Nevatia, R.: Segmentation and recovery of SHGCs from a real intensity image. In: Proceedings Third European Conference on Computer Vision, Vol. 1, pp. 319–330. Stockholm, Sweden, 1994.

[57] Yokoya, N., Kaneta, M., Yamamoto, K.: Recovery of superquadric primitives from a range image using simulated annealing. In: Proc. 11th Int. Conf. on Pattern Recognition, Vol. A, pp. 168-172. IAPR, 1992. IEEE CS.

Dr. F. Solina
Computer Vision Laboratory
Faculty of Electrical Engineering
and Computer Science
University of Ljubljana
Tržaška c. 25
SI-61001 Ljubljana
Slovenia
e-mail: Franc.Solina@Ger.uni-G.si

Computing Suppl. 11, 221–236 (1996)

Theoretical Foundations of Anisotropic Diffusion in Image Processing*

J. Weickert, Kaiserslautern

Abstract

Theoretical Foundations of Anisotropic Diffusion in Image Processing. A frequent problem in low-level vision consists of eliminating noise and small-scale details from an image while still preserving or even enhancing the edge structure. Nonlinear anisotropic diffusion filtering may be one possibility to achieve these goals. The objective of the present paper is to review the author's results on a scale-space interpretation of a class of diffusion filters which comprises also several nonlinear anisotropic models. It is demonstrated that these models—which use an adapted diffusion tensor instead of a scalar diffusivity—offer advantages over isotropic filters. Most of the restoration and scale-space properties carry over from the continuous to the discrete case. Applications are presented ranging from pre-processing of medical images and postprocessing of fluctuating numerical data to visualizing quality relevant features for the grading of wood surfaces and fabrics.

Key words: Anisotropic diffusion, image enhancement, scale-space, maximum principle, Lyapunov functions, CAQ.

1. Introduction

In recent years, nonlinear diffusion methods have proved to be useful in many fields ranging from medical applications [3, 11, 22] and image-sequence analysis [24, 30] to computer aided quality control [33] and postprocessing of noisy data [32].

Nevertheless, there seems to remain both practical and theoretical problems:

- In many practical applications, the first nonlinear diffusion technique due to Perona and Malik [23] is used. This method is well-known to give poor results for very noisy images. But also more robust regularizations may still lead to problems at noisy edges, as long as they use scalar-valued diffusivities [33].
- For most nonlinear diffusion filters, no correct theory is available: questions of existence and uniqueness of a solution are hardly addressed, almost no stability analysis is performed, and concerning the steady-state behaviour, conjectures

* This work was supported by "Stiftung Volkswagenwerk" and "Stiftung Rheinland-Pfalz für Innovation".

dominate the field. As exceptions let us mention the works of Catté et al. [6], Cottet and Germain [8], and Schnörr [29]. As long as these questions are unsolved, there is no way for a convincing scale-space interpretation of non-linear diffusion filtering.

The present paper gives a survey of the author's work to address these problems. To this end, we investigate two regularized, nonlinear anisotropic diffusion filters which use a diffusion tensor instead of a scalar diffusivity and perform well even on very noisy images. We shall see that they belong to a general class of diffusion models, which includes certain previous models and for which a correct theory is available. Finally, we demonstrate their practical use as an image enhancement tool by discussing many examples.

The outline of the paper is as follows: Section 2 describes the essential ideas of anisotropic diffusion filtering and its mathematical legitimation. It is illustrated that the proposed models allow contrast enhancement and offer advantages at noisy edges compared to less sophisticated diffusion techniques. The third section is dedicated to a scale-space interpretation of anisotropic diffusion. We focus especially on the aspects, in which sense a contrast enhancing image processing technique can still be regarded as a smoothing transformation. Section 4 sketches briefly numerical methods for diffusion filtering and in the fifth section, we discuss practical applications of this technique. We conclude with a summary in Section 6.

2. Image Enhancement by Anisotropic Diffusion

Let the image domain be an open rectangle $\Omega := (0, a_1) \times (0, a_2)$, $\Gamma := \partial\Omega$ its boundary and let an image $f(x)$ be represented by a bounded function $f \colon \Omega \to \mathbb{R}$. Then, a filtered version $u(x, t)$ of $f(x)$ with a scale parameter $t \geq 0$ may be obtained as the solution of a diffusion equation with f as initial condition and reflecting boundary conditions:

$$\partial_t u = \mathrm{div}(D(\nabla u_\sigma)\nabla u) \quad \text{on} \quad \Omega \times (0, \infty) \tag{1}$$

$$u(x, 0) = f(x) \quad \text{on} \quad \Omega \tag{2}$$

$$\langle D(\nabla u_\sigma)\nabla u, n \rangle = 0 \quad \text{on} \quad \Gamma \times (0, \infty) \tag{3}$$

Hereby, n denotes the outer normal and $\langle ., . \rangle$ the usual inner product. The symmetric positive definite diffusion tensor $D \in \mathbb{R}^{2 \times 2}$ is chosen to be a function of the edge estimator ∇u_σ, where

$$K_\sigma(x) := \frac{1}{2\pi\sigma^2} \cdot \exp\left(-\frac{|x|^2}{2\sigma^2}\right), \tag{4}$$

$$u_\sigma(x, t) := (K_\sigma * \tilde{u}(., t))(x) \qquad (\sigma > 0) \tag{5}$$

and $\tilde{u}$ denotes an extension of u from Ω to $\mathbb{R}^2$, which may be obtained by mirroring at Γ. The regularization by convolving with a Gaussian makes the edge detection insensitive to noise at scales smaller than σ and helps to ensure existence and uniqueness results in a similar way as in [6].

The preceding filter strategy covers a wide class of diffusion methods, classic ones as well as new ones. To see this, we have to specify the diffusion tensor. Let us study three examples:

(a) *Linear isotropic diffusion filtering*

The simplest diffusion filter utilizes the unit matrix as diffusion tensor:

$$D(\nabla u_\sigma) = I. \tag{6}$$

This technique goes back to Marr and Hildreth [18], Witkin [37], and Koenderink [14]. It is widely used in the image processing community, since it is equivalent to convolving the original image with Gaussians of increasing size (when disregarding the boundary conditions). Nevertheless, it smoothes noise within a region in the same way as it blurs semantically important structures like edges.

(b) *Nonlinear isotropic diffusion filtering*

In order to avoid blurring of edges, one should construct a diffusion filter which reduces the diffusivity at those locations which are good candidates for being an edge. To this end, consider a decreasing function $g \in C^\infty([0,\infty),(0,1])$, which can be represented on $[0,\infty)$ by a convergent power series and satisfies $g(0) = 1$ and $g(s) \to 0$ for $s \to \infty$. For example, one may take

$$g(s) = \exp\left(\frac{-s^5}{5\lambda^5}\right). \tag{7}$$

The theoretically very well-investigated model of Catté, Lions, Morel and Coll [6] utilizes a diffusion tensor of type

$$D(\nabla u_\sigma) = g(|\nabla u_\sigma|)I. \tag{8}$$

This idea boils down to the linear diffusion case (a) in the interior of a region ($|\nabla u_\sigma| \to 0$) and inhibits diffusion at strong edges ($|\nabla u_\sigma| \to \infty$).

However, the inhibition is isotropic, since all eigenvalues of D are reduced by the same amount. Thus, noise at edges cannot be removed by permitting more diffusion along the edge than across it. In order to achieve this, we have to consider anisotropic models.

(c) *Nonlinear anisotropic diffusion filtering*

Anisotropic models do not only take into account the modulus of the edge detector ∇u_σ, but also its direction. We construct the orthonormal system of eigenvectors v_1, v_2 of D such that they reflect the edge structure:

$$v_1 \| \nabla u_\sigma, \qquad v_2 \perp \nabla u_\sigma.$$

In order to prefer smoothing along the edge to smoothing across it, one should choose the corresponding eigenvalues λ_1 and λ_2 such that

$$\frac{\lambda_1(|\nabla u_\sigma|)}{\lambda_2(|\nabla u_\sigma|)} \to 0 \qquad \text{for} \quad |\nabla u_\sigma| \to \infty.$$

As in the linear and nonlinear isotropic case, we may wish to reduce noise within a region by means of eigenvalues with

$$\lim_{s\to 0} \lambda_1(s) = \lim_{s\to 0} \lambda_2(s) = 1.$$

Among the numerous ways to construct such a diffusion tensor, let us investigate two examples which were proposed by Weickert: Let $e_\varphi = (\cos\varphi, \sin\varphi)^T$, and $a \otimes b = ab^T$. Then the integration model [31, 33] is given by

$$D(\nabla u_\sigma) := \frac{2}{\pi} \int_0^\pi e_\varphi \otimes e_\varphi g(|\langle \nabla u_\sigma, e_\varphi\rangle|)\,d\varphi, \tag{9}$$

and the tensor product model [32] utilizes

$$D(\nabla u_\sigma) := g(\sqrt{\nabla u_\sigma \otimes \nabla u_\sigma}). \tag{10}$$

The main difference between (9) and (10) is the smoothing behaviour along edges: for $|\nabla u_\sigma| \to \infty$, (9) gives $\lambda_2(|\nabla u_\sigma|) \to 0$, whereas (10) yields $\lambda_2(|\nabla u_\sigma|) \to 1$. Thus, the second proposal is stronger anisotropic. This is not always an advantage: in some cases, it may also lead to a slightly stronger rounding of corners [32].

Besides the preceding examples, other anisotropic diffusion models can be found in the literature, see [8, 9, 20]. A related anisotropic image restoration method may also be constructed using the mean-curvature equation [2].

The filter class (1)–(3) with diffusion tensors such as (6), (8), (9), (10) is mathematically sound, as it possesses a unique solution in the distributional sense, which is infinitely often differentiable for $t > 0$. This distinguishes these filters from nonlinear diffusion techniques such as [23] and [21], which are claimed to be ill-posed [6]. The proof for the case (b) may be found in [6]. It includes (a) and can be extended in a straightforward way to case (c). For a more general characterization of the class of diffusion filters which satisfy these properties, see [34].

Let us now compare the image restoration properties of the examples (a)–(c). Figure 1a consists of a triangle and a rectangle with 70% of all pixel being completely degraded by noise. It is taken from the Software package *MegaWave* which was developed at the CEREMADE (University Paris IX). All images in the present paper possess a range within the interval $[0, 255]$, and they are depicted in such a way that the lowest value is black and the highest one appears white. In Fig. 1b we observe that linear diffusion filtering is capable of removing all noise, but we have to pay a price: the image becomes completely blurred. Besides the fact that edges get smoothed so that they are harder to identify, a second problem appears, the

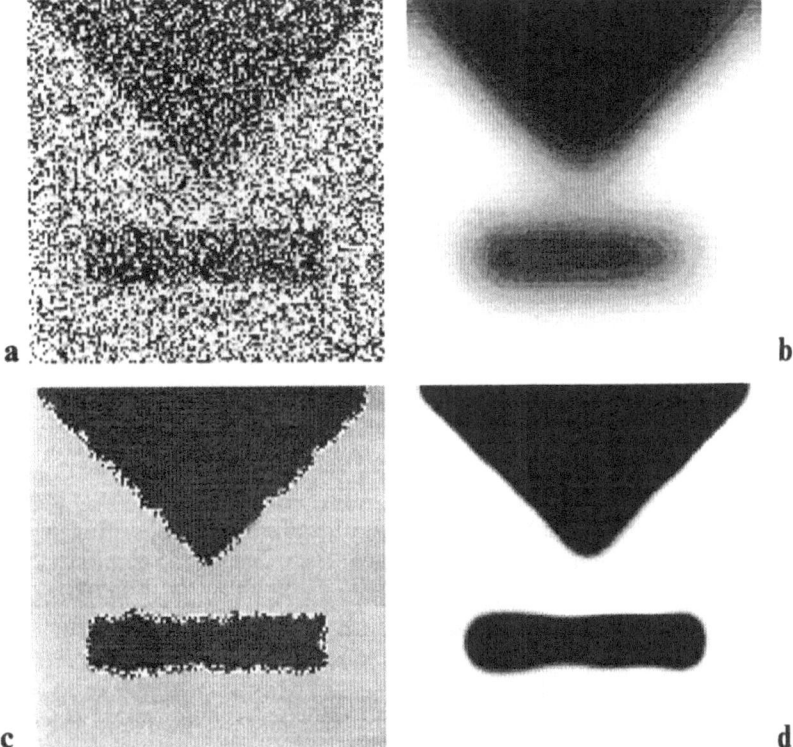

Figure 1. Comparison of different diffusion filters. **a** Test image, $\Omega = (0,127)^2$. **b** Linear diffusion, $t = 80$. **c** Nonlinear isotropic diffusion, $\lambda = 3.5$, $\sigma = 3$, $t = 80$. **d** Nonlinear anisotropic diffusion, tensor product model, $\lambda = 3.5$, $\sigma = 3$, $t = 80$

so-called correspondence problem: edges become more and more dislocated. Thus, once they are identified at a coarse scale, they have to be traced back in order to find their true location, a numerically very difficult problem.

Nonlinear isotropic diffusion does not show these correspondence problems, since edges are hardly affected by the process. On the other hand, they are actually too less affected: noise at edges remains, as it is to be seen in Fig. 1c.

Figure 1d demonstrates that nonlinear anisotropic filtering shares the advantages of both beforementioned methods. It combines good noise eliminating properties of linear diffusion with the stable edge structure of nonlinear isotropic diffusion.

Anisotropic diffusion filtering is an image restoration tool that needs essentially two natural parameters: a contrast parameter λ and a resolution parameter σ. Applying the same reasoning as Catté et al. [6], we may relate the stopping time T to σ by choosing $T = 0(\sigma^2)$.

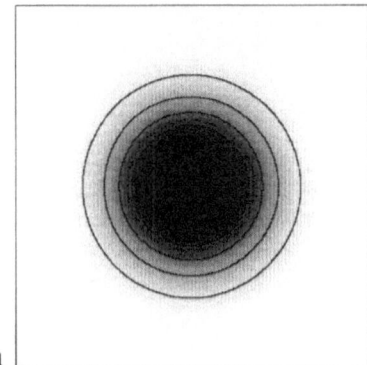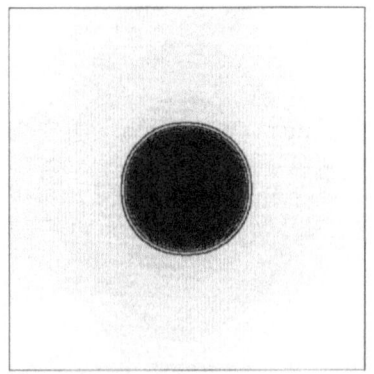

a ⎵⎵⎵⎵⎵⎵⎵⎵⎵⎵⎵⎵⎵⎵⎵⎵⎵⎵⎵⎵⎵⎵⎵⎵⎵⎵⎵⎵⎵⎵⎵⎵ **b**

Figure 2. Edge enhancement of anisotropic diffusion. **a** Gaussian-type function, $\Omega = (0,255)^2$. **b** Filtered, tensor product model, $\lambda = 3, \sigma = 2, t = 1800$

For fast decreasing diffusivities such as (7), an interesting behaviour can be observed: since the diffusion at both sides of an edge is much stronger than at the edge itself, one can investigate contrast enhancement at the edge. Hence, although the diffusivity is always nonnegative, the filter behaves like backward diffusion at edges and like forward diffusion within a region. A more detailed analysis of this phenomenon can be found in [23]. Figure 2 illustrates contrast enhancement using anisotropic diffusion filtering. It depicts a 2D Gaussian-like function with its isolines before and after processing. It can be observed that two regions with almost constant grey value evolve which are separated by a fairly steep edge. The segmentation-like results in Figs. 1d and 2b indicate also that nonlinear diffusion filtering is a useful preprocessing tool making subsequent segmentations very easy.

3. Scale-Space Properties

Besides its qualities as an image enhancing tool, anisotropic diffusion filtering can also be regarded as a smoothing transformation which gives a scale-space representation of the original image. In the following, we will denote by $u(x, t)$ the unique solution of the filter class (1)–(3) with diffusion tensors such as (6), (8), (9) or (10).

3.1. The Scale-Space Concept

In order to get an impression of the concept of scale-spaces, let us recall that images usually contain structures at a large variety of scales. In those cases where it is not clear in advance which is the right scale for the depicted information, it is desirable to represent the image at multiple scales. Moreover, by comparing the structures at different scales, one obtains a hierarchy of image structures, which eases a subsequent image interpretation.

A scale-space is a representation at a continuum of scales, embedding the original image f into a family $\{T_t f \mid t \geq 0\}$ of gradually simplified versions of it. A scale-space representation has to fulfil several architectural, smoothing (information reducing) and invariance requirements [1], which we will briefly review now:

The semigroup property is a typical representative for an architectural requirement. It states that for $t = 0$, the scale-space representation gives the original image f, and the filtering may be split into a sequence of filter banks:

$$T_0 f = f,$$

$$T_{t+s} f = T_t (T_s f) \qquad \forall s, t \geq 0.$$

Information reduction arises from the wish that the smoothing transformation should not create artefacts when passing from fine to coarse representation. Thus, at a coarse scale, we must not have additional structures which are caused by the filtering method itself and not by underlying structures at finer scales.

It is desirable that—except for information reduction by smoothing—the scale-space operator does not alter the image too much, i.e. it should be invariant under many transformations. Examples of such transformations are grey-level shifts, translations and rotations, but also affine transformations of the space.

The work of Alvarez, Guichard, Lions and Morel [1] shows that every scale-space fulfilling some fairly natural architectural, information reducing and invariance properties is governed by a partial differential equation (PDE) with the original image as initial condition.

Examples for scale-space generating PDEs are the linear diffusion equation [37, 14]

$$\partial_t u = \Delta u \tag{11}$$

and the affine invariant morphological equation [1, 28]

$$\partial_t u = |\nabla u| \left(\operatorname{div} \left(\frac{\nabla u}{|\nabla u|} \right) \right)^{1/3}. \tag{12}$$

Morphological transformations possess the property that the filtering result depends only on the level sets of the image and, therefore, they are invariant under any nondecreasing grey level transformation (*grey scale invariance*) [1]. On the other hand, it is evident that a grey scale invariance requirement is not compatible with any contrast dependent image enhancement method. Hence, if one insists in having scale-spaces that allow contrast enhancement (like in our case), one has to withdraw morphology.

The goal of this section is to establish the proposed filter class as scale-space transformations. To this end, we shall not focus on further investigations of architectural requirements, as these qualities are automatically fulfilled and do not distinguish nonlinear diffusion scale-spaces from other ones. We start with briefly discussing invariances. Afterwards, we turn to a more crucial task, namely the question, in which sense our restoration method—which allows edge enhancement —can still be considered as a smoothing, information reducing image transformation. As this paper is intended to give a concise survey on this subject, we will focus on the main ideas. The reader who is interested in the proofs and mathematical details is referred to a technical report by the author [34].

3.2. Invariances

By the construction of the proposed filters, it is not hard to verify that they are invariant under grey level shifts and contrast reversions.

If one omits boundary conditions and regards diffusion filtering as a pure initial value problem on an infinite domain, it makes also sense to consider translations and rotations of the image. In this case, anisotropic diffusion filtering is also invariant under translations and isometric transformations.

For other invariances, the homogeneous Neumann boundary conditions (3) are very useful: together with the divergence form of the diffusion equation, they imply that the average grey level

$$\mu := \frac{1}{|\Omega|} \int_\Omega f(x)\,dx \tag{13}$$

is not affected by nonlinear diffusion filtering:

$$\frac{1}{|\Omega|} \int_\Omega u(x,t)\,dx = \mu \qquad \forall t > 0.$$

This distinguishes diffusion scale-spaces from morphological ones. In general, the latter ones are not of divergence form and do not preserve the average grey level.

3.3. Information Reducing Properties

3.3.1. Nonenhancement of Local Extrema

The requirement that a scale-space representation must not amplify local extrema was first pointed out by Lindeberg [16] for the linear diffusion scale-space. However, this condition is also satisfied by nonlinear anisotropic diffusion [34]. Let us consider an arbitrary, but fixed time $\theta > 0$ and suppose that $\xi \in \Omega$ is a local extremum of $u(.,\theta)$. Then,

$$\partial_t u(\xi,\theta) \le 0, \qquad \text{if } \xi \text{ is a local maximum,}$$

$$\partial_t u(\xi,\theta) \ge 0, \qquad \text{if } \xi \text{ is a local minimum.}$$

Nonenhancement of local extrema contrasts anisotropic diffusion to classical image enhancing methods such as high-frequency emphasis ([12], pp. 182–183), which do violate this principle. Although possibly being in the backward diffusion region at edges, nonlinear diffusion is always in the forward region at extrema. This ensures its stability.

3.3.2. Maximum-Minimum Principle and Consequences

An extremum principle is a property which is closely related to the nonenhancement of local extrema. A common way to prove extremum principles for nonlinear parabolic PDEs is the use of monotony results (see e.g. [25], pp. 186–188). The monotony condition

$$f \leq h \Rightarrow T_t f \leq T_t h \qquad \forall t > 0.$$

has also been proposed as a smoothing requirement for scale-spaces [1]. When combined with reverse contrast invariance and grey level shift invariance, this comparison principle implies an extremum principle. However, to establish the monotony condition for anisotropic diffusion filtering, we would need that the equation is always of forward parabolic type. Since this would forbid contrast enhancing processes, we will not pursue this idea any further.

But we are not lost: In order to prove a maximum-minimum principle, we may utilize a technique which does not require monotony, namely Stampacchia's truncation method (cf. [4], p. 211). Suppose that the original image f is bounded by some $a, b \in \mathbb{R}$:

$$a \leq f(x) \leq b \qquad \forall x \in \Omega.$$

Then, Stampacchia's truncation method implies that the filtered image remains within these bounds for all times [34]:

$$a \leq u(x, t) \leq b \quad \text{on} \quad \bar{\Omega} \times (0, \infty). \tag{14}$$

This result is of essential practical importance, as it guarantees for instance that if we start with an image within a range $[0, 255]$, we will never get a result with a grey value such as 257.

Hummel [13] shows that, under certain conditions, the extremum principle for parabolic operators is equivalent to the property that the corresponding scale-space never creates additional level-crossings for $t > 0$. This points out the importance of extremum principles for scale-spaces.

Furthermore, the preceding maximum-minimum principle is a useful tool for establishing theoretical properties like continuous dependence of the solution on the initial data. This is an important step for proving the well-posedness of our method.

We denote by $L^p(\Omega)$ with $1 \leq p < \infty$ the space of functions $f: \Omega \to \mathbb{R}$, for which $|f|^p$ is Lebesgue integrable, and provide it with the norm

$$\|f\|_{L^p(\Omega)} := \left(\int_\Omega |f(x)|^p \, dx \right)^{1/p}.$$

Then it can be shown that the solution $u(x, t)$ of the diffusion filter depends continuously on the initial image f with respect to the $L^2(\Omega)$ norm [34]. This result is especially important when applying anisotropic diffusion in areas like stereo vision or analysis of image sequences, as it guarantees that similar images remain similar after being processed. Although this requirement is hardly mentioned in scale-space theory, it deserves to be carefully checked. A scale-space violating this property can be practically completely unstable in spite of having plenty of smoothing and invariance qualities.

3.3.3. Behaviour for $t \to \infty$

As scale-spaces are intended to simplify an image, it would be desirable that, for $t \to \infty$, the result converges to the simplest possible image representation, namely a constant image with the same average grey value as the original one. Unlike morphological equations such as (12), which may have nontrivial steady-states, our diffusion filter class satisfies this requirement [34]. The convergence to the constant image is guaranteed in every L^p norm with $1 \leq p < \infty$.

3.3.4. Lyapunov Functionals

We have seen that, for $t \to \infty$, the diffusion filtered image becomes completely smooth. On the other hand, we observed that for finite times, locally the opposite may appear: contrast can be enhanced. Hence the question arises, whether there exists some quantity which indicates that the image becomes steadily smoother from a global view.

We have already found such global quantities: The extremum principle tells us that the global minimum increases and the global maximum decreases with respect to time. But there exist numerous other quantities as well [34]: all $L^p(\Omega)$ norms of $u(x, t)$ with $2 \leq p < \infty$ are decreasing in t. This comprises also the energy $\|u(t)\|^2_{L^2(\Omega)}$.

Another class of decreasing global qualities is given by the even central moments [34]

$$M_{2n}[u(t)] := \frac{1}{|\Omega|} \int_\Omega (u(x, t) - \mu)^{2n} \, dx \qquad (n \in \mathbb{N}).$$

The second central moment (the variance) characterizes the spread of the intensity about its mean. It is a common tool for constructing measures for the relative smoothness of the intensity distribution. The fourth moment is frequently used to describe the relative flatness of the grey value distribution. Higher moments are more difficult to interprete, although they do provide important information for tasks like texture discrimination ([12], pp. 414–415). All decreasing even moments demonstrate that the image becomes smoother during diffusion filtering. Hence, local effects such as edge enhancement, which object to increase central moments, are overcompensated by strong smoothing in other areas.

If the initial image f is strictly positive on Ω, we may regard it also as a two-dimensional density. (Without loss of generality, we omit the normalization.) Then,

$$S[u(t)] := - \int_{\Omega} u(x, t) \ln(u(x, t)) \, dx$$

is called the entropy of $u(t)$, a measure of uncertainty and missing information [5]. Anisotropic diffusion filtering can be shown to increase the entropy [34], hence the corresponding scale-space embeds the genuine image f into a family of subsequently likelier versions of it which contain less information. Moreover, for $t \to \infty$, the process reaches the state with the lowest possible information, namely a constant image.

From all the previous considerations, we observe that, in spite of its edge enhancing properties, anisotropic diffusion does really simplify the original image in a steady way.

4. Numerical Approximation

Since a scale-space representation or an enhancement method cannot be better than its numerical realization, it is of crucial importance to study the discrete case of diffusion filtering.

For diffusion problems, numerous numerical methods can be applied:

In [10], a comparison was made between three schemes for a one-dimensional model of nonlinear diffusion filtering: a wavelet method of Petrov-Galerkin type, a spectral method and a finite-difference (FD) scheme. It turned out that—especially for large σ—all results were fairly similar. Since the computational effort is of a comparable order of magnitude, it seems to be a matter of taste which scheme one prefers. Of course, other numerical methods are applicable as well, e.g. finite elements. Neural network realizations of nonlinear diffusion filters were proposed by Cottet [7, 9] and a multigrid acceleration for a scheme which is related to nonlinear diffusion filtering was studied by Saint-Marc, Chen and Medioni [27]. However, most nonlinear diffusion researchers prefer finite differences, since they are easy to handle and the pixel structure of a real digital image provides already a natural discretization on a fixed rectangular grid. Thus, in the sequel we restrict our considerations to the FD case.

Besides low complexity, the requirements for a good numerical scheme for diffusion in image processing differ from requirements in other areas such as computational fluid dynamics: very accurate high-order approximations of the continuous equation are less important than approximations which exhibit as many qualitative properties of the continuous equation as possible. So the question arises which of the continuous scale-space properties carry over to the FD case.

It is not hard to see that properties like grey level shift invariance or reverse contrast invariance are easily fulfilled by every reasonable FD scheme. Translation invariance and isometry invariance seem to be more problematic, since translations make only sense for multiples of the grid distance, and isometry invariance can only be approximately satisfied.

In [35], it is investigated under which requirements one can construct semidiscrete (i.e. continuous in time, discrete in space) and fully discrete anisotropic diffusion filters with the following properties: well-posedness, average grey level invariance, maximum-minimum principle, existence of Lyapunov functionals, and convergence to a constant steady-state. It will be shown that it is possible to find consistent FD schemes having all these qualities. These comprise also splitting-based, absolutely stable semiimplicit schemes requiring per time step a computational and storage effort which is linear in the pixel number.

5. Applications

At the Laboratory of Technomathematics, diffusion filtering is currently applied in the following areas.

(a) *Postprocessing of data with numerical fluctuations* [31]
   Particle methods for solving the Boltzmann equation utilize random or pseudorandom processes which lead to numerical fluctuations. By means of nonlinear diffusion filtering it was possible to smooth "jittery" isolines without affecting their principal structure. Meanwhile, nonlinear diffusion is used as a tool to process the Boltzmann solution in such a way that it can be coupled with the smooth solution of the Navier-Stokes equation in adjacent regions. [17].

(b) *Grading of fabrics* [33]
   The quality of a fabric is determined by two criteria, namely clouds and stripes. Clouds result from isotropic inhomogeneities of the density distribution, whereas stripes are an anisotropic phenomenon caused by adjacent fibers pointing in the same direction. Anisotropic diffusion filters are capable of visualizing both quality-relevant features simultaneously (Fig. 3). For a suitable parameter choice, they smooth in an isotropic way at clouds and diffuse anisotropically along fibers in order to enhance them.

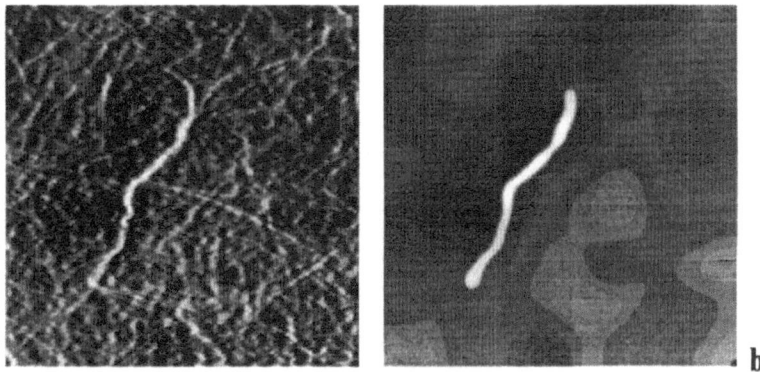

Figure 3. Preprocessing of fabric images. **a** Fabric, $\Omega = (0,256)^2$. **b** Filtered, tensor product model, $\lambda = 4, \sigma = 2, t = 160$

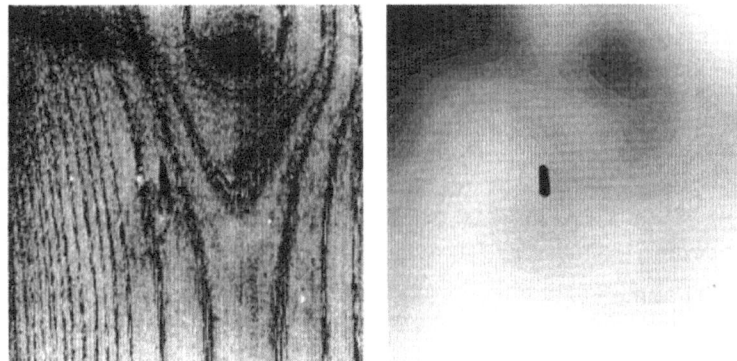

Figure 4. Defect detection in wood. **a** Wood surface, $\Omega = (0,255)^2$. **b** Filtered, integration model, $\lambda = 4$, $\sigma = 2.8, t = 320$

(c) *Defect detection of wood surfaces* [31, 33]

For furniture production it is of importance to classify the quality of wood surfaces. If one aims to automize this evaluation, one has to process the image in such a way that quality relevant features become better visible und unimportant structures disappear. Figure 4a depicts a wood surface possessing one defect. By means of anisotropic diffusion filtering it is possible to visualize this fault in a much better way (Fig. 4b). In [33] it is demonstrated, how a further modification of the diffusion tensor yields even more accurate results with less roundings at the defect's corners.

(d) *Preprocessing of medical images*

Figure 5a gives an example for possible medical applications of anisotropic diffusion. It depicts an MR image of the liver. To ease the diagnosis it is intended to apply segmentation algorithms for classifying the different types of tissue. Figure 5c shows a segmentation according to the Mumford-Shah energy

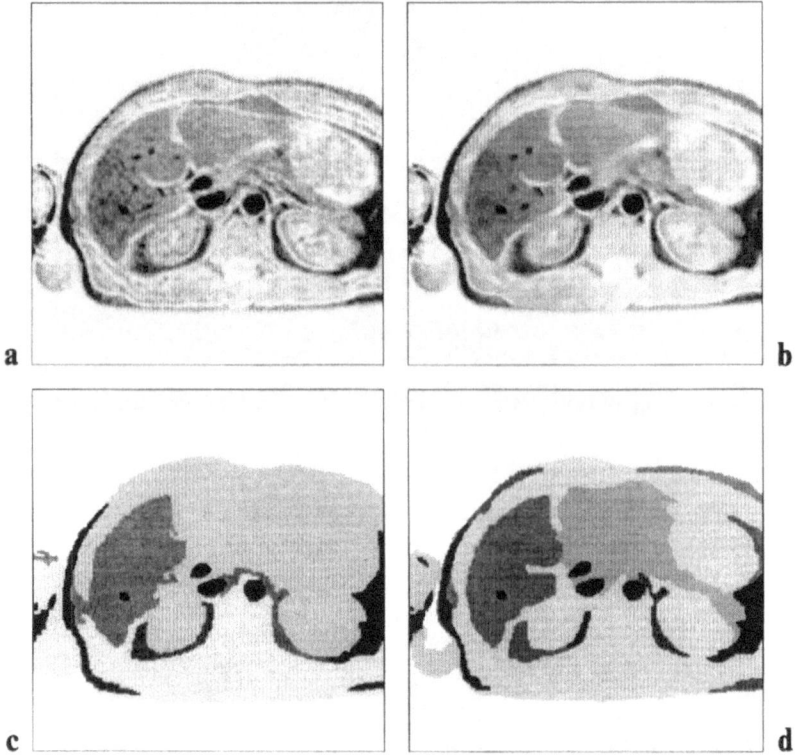

Figure 5. Preprocessing of MR images. **a** Liver, $\Omega = (0,255)^2$. **b** Filtered, integration model, $\lambda = 1$, $\sigma = 3, t = 4$. **c** Segmented original image. **d** Segmented filtered image

functional [19] by means of an algorithm due to Koepfler, Morel and Solimini [15]. As it is seen in Fig. 5d, one can obtain more realistic results when processing the original image by means of anisotropic diffusion (Fig. 5b) prior to segmentation. An algorithm for 3D images is currently under investigation, which allows to filter simultaneously an entire MRI or CT set [26]. To this end, a version for parallel machines of MIMD type was developed.

6. Conclusions

We have investigated anisotropic diffusion filters which use an adapted diffusion tensor instead of a scalar diffusivity. This quality distinguishes them from most other nonuniform diffusion methods such as [6, 21–23, 29, 36] and offers advantages at noisy edges. The proposed filters belong to a class which is theoretically well-founded, as it permits existence, uniqueness and regularity results, continuous dependence of the solution on the initial image, a maximum-minimum principle, and convergence to a constant steady-state as t tends to infinity.

Anisotropic diffusion filtering is capable of enhancing images by smoothing isotropically within a region, while diffusing in an anisotropic way along edges. Rapidly decreasing diffusivities allow even contrast enhancement at edges.

In spite of this property, anisotropic nonlinear diffusion filters turn out to be smoothing with respect to numerous aspects. Together with certain architectural and invariance properties, they induce a scale-space representation of the original image, which may incorporate user-specific demands on contrast and size of especially interesting structures. Edges remain well-pronounced and stable across a wide range of scales.

The presented examples demonstrate that anisotropic diffusion is well-suited as a pre- and postprocessing tool and as a method to visualize quality relevant features in the area of computer aided quality control (CAQ).

Furthermore, it should be mentioned that the previous results can be easily extended to higher dimensions, vector-valued images and more sophisticated descriptors of local structure allowing also corner enhancement [34].

References

[1] Alvarez, L., Guichard, F., Lions, P.-L., Morel, J.-M.: Axioms and fundamental equations in image processing. Arch. Rational Mech. Anal. *123*, 199–257 (1993).
[2] Alvarez, L., Lions, P.-L., Morel, J.-M.: Image selective smoothing and edge detection by nonlinear diffusion. II. SIAM J. Numer. Anal. *29*, 845–866 (1992).
[3] Bajla, I., Marušiak, M., Šrámek, M.: Anisotropic filtering of MRI data based upon image gradient histogram. In: Computer analysis of images and patterns (Chetverikov, D., Kropatsch, W. G., eds.), pp. 90–97. Berlin, Heidelberg, New York, Tokyo: Springer 1993.
[4] Brezis, H.: Analyse fonctionelle. Paris: Masson, 1983.
[5] Buck, B., Macaulay, V. (eds.): Maximum entropy in action. Oxford: Clarendon 1991.
[6] Catté, F., Lions, P.-L., Morel, J.-M., Coll, T.: Image selective smoothing and edge detection by nonlinear diffusion. SIAM J. Numer. Anal. *29*, 182–193 (1992).
[7] Cottet, G.-H.: Diffusion approximation on neural networks and applications for image processing. In: Proc. Sixth European Conference on Mathematics in Industry (Hodnett, F., ed.), pp. 3–9. Stuttgart: Teubner 1992.
[8] Cottet, G.-H., Germain, L.: Image processing through reaction combined with nonlinear diffusion. Math. Comp. *61*, 659–673 (1993).
[9] Cottet, G.-H.: Neural networks: Continuous approach and applications to image processing. Technical Report No. 113, LMC—IMAG, Université Joseph Fourier, B.P. 53, 38041 Grenoble Cédex 9, France, 1994 (submitted to Proc. 2nd Europ. Conf. on Mathematics applied to Biology and Medicine, Lyon, Dec. 15–18, 1993).
[10] Fröhlich, J., Weickert, J.: Image processing using a wavelet algorithm for nonlinear diffusion. Report No. 104, Laboratory of Technomathematics, University of Kaiserslautern, P.O. Box 3049, 67653 Kaiserslautern, Germany, 1994.
[11] Gerig, G., Kübler, O., Kikinis, R., Jolesz, F. A.: Nonlinear anisotropic filtering of MRI data. IEEE Trans. Medical Imaging *11*, 221–232 (1992).
[12] Gonzalez, R. C., Wintz, P.: Digital image processing. Reading: Addison-Wesley 1987.
[13] Hummel, R. A.: Representations based on zero-crossings in scale space. Proc. IEEE Computer Society Conf. on Comp. Vision and Pattern Recognition, pp. 204–209, 1986.
[14] Koenderink, J. J.: The structure of images. Biol. Cybern. *50*, 363–370 (1984).
[15] Koepfler, G., Morel, J.-M., Solimini, S.: Segmentation by minimizing a functional and the "merging" method. Preprint No. 9022, CEREMADE, Université Paris IX—Dauphine, Place du Maréchal de Lattre de Tassigny, 75775 Paris Cédex 16, France, 1990.

[16] Lindeberg, T.: Scale-space for discrete signals. IEEE Trans. Pattern Anal. Mach. Intell. *12*, 234–254 (1990).

[17] Lukschin, A., Neunzert, H., Struckmeier, J.: Interim report of the project DPH 6473/91—Coupling of Navier-Stokes and Boltzmann regions. Internal Report, Kaiserslautern, 1993.

[18] Marr, D., Hildreth, E.: Theory of edge detection. Proc. R. Soc. London Ser. *B 207*, 187–217 (1980).

[19] Mumford, D., Shah, J.: Optimal approximations by piecewise smooth functions and associated variational problems. Comm. Pure Appl. Math. *42*, 577–685 (1989).

[20] Nitzberg, M., Shiota, T.: Nonlinear image filtering with edge and corner enhancement. IEEE Trans. Pattern Anal. Mach. Intell. *14*, 826–833 (1992).

[21] Nordström, N.: Biased anisotropic diffusion—a unified regularization and diffusion approach to edge detection. Image Vision Comput. *8*, 318–327 (1990).

[22] Ottenberg, K.: Model-based extraction of geometric structure from digital images. Ph.D. thesis, Utrecht University, The Netherlands, 1993.

[23] Perona, P., Malik, J.: Scale space and edge detection using anisotropic diffusion. IEEE Trans. Pattern Anal. Mach. Intell. *12*, 629–639 (1990).

[24] Proesmans, M., Van Gool, L., Pauwels, E., Oosterlinck, A.: Determination of optical flow and its discontinuities using non-linear diffusion. In: Computer Vision—ECCV '94 (Eklundh, J.-O., ed.), pp. 295–304. Berlin, Heidelberg, New York, Tokyo: Springer 1994 (Lecture Notes in Computer Science, Vol. 801).

[25] Protter, M. H., Weinberger, H. F.: Maximum principles in differential equations. Englewood Cliffs: Prentice-Hall 1978.

[26] Rambaux, I., Garçon, P.: Nonlinear anisotropic diffusion filtering of 3D images. Project Work, Département Génie Mathématique, INSA de Rouen and Laboratory of Technomathematics, University of Kaiserslautern, 1994.

[27] Saint-Marc, P., Chen, J. S., Medioni, G.: Adaptive smoothing: a general tool for early vision. IEEE Trans. Pattern Anal. Mach. Intell. *13*, 514–529 (1990).

[28] Sapiro, G., Tannenbaum, A.: Affine invariant scale-space. Int. J. Comput. Vision *11*, 25–44 (1993).

[29] Schnörr, C.: Unique reconstruction of piecewise smooth images by minimizing strictly convex non-quadratic functionals. J. Math. Imag. Vision *4*, 189–198 (1994).

[30] Schnörr, C.: Bewegungssegmentation von Bildfolgen durch die Minimierung konvexer nicht-quadratischer Funktionale. In: Tagungsband Mustererkennung 1994 (Kropatsch, W. G., Bischof, H., eds.), pp. 178–185. Informatik Xpress 5, Wien, 1994.

[31] Weickert, J.: Zwischenbericht zum Projekt "Nichtlineare Diffusionsfilter". Bericht über die wissenschaftliche Tätigkeit Januar 1991–Dezember 1991, Center for Applied Mathematics, Darmstadt–Kaiserslautern, pp. 133–142, 1992.

[32] Weickert, J.: Abschlußbericht zum Projekt "Nichtlineare Diffusionsfilter". Abschlußbericht und Bericht über die wissenschaftliche Tätigkeit Januar 1992–Dezember 1993, Center for Applied Mathematics, Darmstadt–Kaiserslautern, pp. 191–209, 1994.

[33] Weickert, J.: Anisotropic diffusion filters for image processing based quality control. In: Proc. Seventh European Conf. on Mathematics in Industry (Fasano, A., Primicerio, M., eds.), pp. 355–362. Stuttgart: Teubner 1994.

[34] Weickert, J.: Scale-space properties of nonlinear diffusion filtering with a diffusion tensor. Report No. 110, Laboratory of Technomathematics, University of Kaiserslautern, P.O. Box 3049, 67653 Kaiserslautern, Germany, 1994 (submitted).

[35] Weickert, J.: Anisotropic diffusion in image processing. Ph.D. thesis, Kaiserslautern, 1995 (to be filed).

[36] Whitaker, R. T., Pizer, S. M.: A multi-scale approach to nonuniform diffusion. CVGIP: Image Understanding *57*, 99–110 (1993).

[37] Witkin, A. P.: Scale-space filtering. Proc. Eighth Int. Joint Conf. on Artificial Intell., Karlsruhe, pp. 1019–1022, 1983.

J. Weickert
Laboratory of Technomathematics
University of Kaiserslautern
P.O. Box 3049
D-67653 Kaiserslautern
Federal Republic of Germany
e-mail: weickert@mathematik.uni-kl.de

Computing Suppl. 11, 237–256 (1996)

© Springer-Verlag 1996

Stability and Likelihood of Views of Three Dimensional Objects

D. Weinshall, M. Werman, and **N. Tishby,** Jerusalem

Abstract

Stability and Likelihood of Views of Three Dimensional Objects. Evaluating the representative power of two dimensional images of three dimensional objects has come up in a number of applications, mostly concerning 3D object recognition. We address the question of image characterization independently of these applications. We first introduce the basic concepts of view stability and likelihood for general objects. The stability and likelihood functions are then used to quantitatively characterize the representability of images. For objects composed of localized features, we develop explicit expressions through which the stability and likelihood functions of any view of a general object can be evaluated from its three principal second moments. This permits a quantitative characterization of the viewing sphere of objects. By way of qualitative analysis we compute the most stable and most likely views, which can identify the characteristic views of objects. We show that the most stable and most likely views of an object are the same and are often unique. This view is the "flattest" view of the object, obtained when the three dimensional object has its minimal spread along the viewing direction. We demonstrate these results using images of familiar objects.

Key words: Quasi-invariance, view-likelihood, view-stability, generic views, canonical views, Bayesian recognition.

1. Introduction

One of the main difficulties in model-based object recognition of three dimensional objects is the geometric source of variability: the object is projected to different two dimensional images when seen from different viewpoints. Often, therefore, the model-based recognition of three dimensional objects from two dimensional images is defined as an indexing and verification process, where for a given model and image the task is to find whether the image (or part of it) can possibly be a projection of the model. This creates the need to develop geometric understanding of the relation between 3D objects and their views.

Model-based recognition is not only about search and verification, however. Often a given image can be the projection of a number of objects (and this is typically the case given noisy images), and the need to choose a single interpretation for the image recasts the model-based object recognition into a statistical framework (as emphasized in the pattern recognition literature). Object recognition becomes a likelihood estimation problem: 'for each object and image, compute the likelihood that the image (or part of it) is the projection of the object from some viewpoint. Choose the object that maximizes this conditional likelihood.'

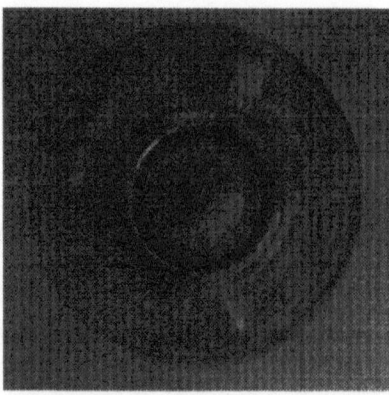

Figure 1. A non generic (unlikely) view of a familiar object, most often interpreted as the image of a glass saucer or a Frisby disc

A typical example for the need in a probabilistic approach to object recognition is an image obtained from a "non-generic" viewpoint, like the one given in Fig. 1. Most people do not recognize on first sight the object photographed in Fig. 1; they do much better using more "generic" views of the same object, such as those in Fig. 2. The likelihood estimation approach can be used to quantify the notion of "generic" views[1], by measuring the conditional probability of the model given the image *Prob(model|image)*. Using Bayes rule, we first rewrite this conditional probability as

$$\text{Prob(model|image)} = \text{Prob(image|model)} \frac{P_m}{P_i}$$

where P_m and P_i denote the prior probabilities of the model and image, respectively. Since P_i is just a normalization integral, in order to select the most likely model we need to maximize the numerator, which contains 2 elements: a prior distribution defined over all models and a conditional distribution of images given the model.

It follows that object recognition, using noisy images and large databases, requires the knowledge of the conditional distribution of images given models, in addition to a prior distribution of models[2]. Although the statistical approach to model-based object recognition has been discussed in the literature in various forms (see Section 1.1), the conditional likelihood *Prob(image|model)* was not derived for general objects and in closed-form. In this paper we take a first step in this

---

[1] For convenience of notations, from now on we will use the term "view" to refer interchangeably to a viewpoint and the image obtained from that viewpoint.

[2] This conditional probability is very small for the above image of a water bottle shown in Fig. 1, and presumably for this reason most people interpret it wrongly. Notice that although an image corresponds to a single view of an object, regardless of the object which induced it, and the conditional probability of any single view is always 0, its *probability density function* depends on the 3D structure of the object and is *positive* for a any possible image of that object.

Figure 2. *Left* A not very likely view of an object, which looks like a lamp shade; *right* a likely view of the object, which is a water bottle

direction, by computing and analyzing the behavior of this function for objects composed of localized features, and by showing several interesting and rather general properties of this function.

Another common way to quantify the notion of "generic" views is via a measure of stability, or how much the image changes as the viewpoint of the object is perturbed (see review of previous work next). This measure does not depend on a given prior probability or the distribution of viewpoints given an object. The idea is that the less stable an image is to small perturbations of the viewing position, the less robust the recognition of the object from the image is. Consequently, in order to obtain a robust recognition system, we would prefer models leading to stable images over models leading to non-stable images, when interpreting a given image.

Goal: in this paper we address the general questions of (conditional) view stability and likelihood. Earlier work analysed specific objects by numerically computing their likelihood function; other papers proposed various stability definitions (see Section 1.1). Our goal here is to define the basic measures of stability and likelihood in the same general framework, and to study and characterize general properties of these measures.

Findings: (1) given objects composed of localized features, we provide explicit expressions that enable us to compute the stability and likelihood of images of any object, using only the 3 principal second moments of the object. (2) By way of qualitative analysis we compute the most stable and most likely views of an object, showing that the most stable and most likely views are the **same** and often **unique**. This view is the "flattest" view of the object, when the three dimensional object has its minimal spread along the viewing direction.

Thus we show that several interesting questions of likelihood and stability can be answered in a very general way, regardless of the particular object of interest.

Outline: the rest of this paper is organized as follows: We develop the concepts of stability and likelihood more precisely in Section 2. In Section 3 we set the background by discussing the comparison of two dimensional images. In Section 4 we develop the expressions for image stability and likelihood of a general object composed of localized features. In Section 5 we discuss a qualitative analysis of image stability and likelihood, which identifies the most-stable and the most-likely view at each aspect. We then demonstrate our results on familiar objects.

Implication: the theory developed here addresses image characterization by likelihood and stability. We foresee its application mostly to object recognition, as described next. We pursue one of these applications in our paper, the quantitative characterization of an aspect in the aspect graph of a pyramid (Section 4.3). This theory is also motivated by and related to human perception, and some of the results reported here can be used to reinterpret psychophysical findings.

1.1. Previous Work: Review and Comparison

The three dimensional structure of objects can be represented in two fundamentally different ways: a two dimensional viewer-centered description, or a three dimensional object-centered description. In a viewer-centered description three dimensional information is not represented explicitly. In employing this approach, an object is represented by a list of 2D characteristic views, that were possibly acquired during a familiarization period. A novel view of the object is recognized by comparing it to the stored views.

The viewer-centered approach is frequently assumed to underlie internal representation in human vision [6, 22, 25]. Recently, computational models were proposed for the computation of intermediate views, using a two layer neural network [10], Radial Basis Functions [23], or linear combinations [27]. Characteristic two dimensional views were used for the representation of three dimensional shape in a number of computational studies (e.g., [8]).

The concept of characteristic views can prove useful for another, object-centered, computational approach to recognition using indexing (e.g. [19]). In this approach, the three dimensional structure of objects is represented by quantities that are invariant to changes in viewing position with respect to the object, and which can be computed directly from an image. Such projective invariants, however, don't generally exist [7, 20], and therefore most of the research in this area has concentrated on the identification of computable invariants for special cases, such as planar objects [11, 17, 20, 28]. Alternatively, the identification of a set of characteristic 2D images, such that any other image of the object is not too far from at least one image in this set, makes it possible to attach to each object a list of "almost" invariant indices, which are computed from the characteristic images.

The concept of "almost" invariance, describing measures that change little over many images, is closely related to the concepts of stability and likelihood discussed above. The attempts to understand and formalize this concept motivated a number of studies:

Likelihood: image likelihood was addressed using numerical simulations of some specific planar angles by Ben-Arie [3] and Burns et al. [7]. Dickinson et al. [9] empirically found the more likely views of some specific objects decomposed into geons.

Stability: Binford [4] defined quasi-invariants, or the local minima of the change in the image when changing the viewing parameters. Other studies proposed to measure stability via the Lie derivatives of the group of transformations describing the motion of the camera [16].

In this earlier work, the analysis of likelihood was carried out for specific objects only (mostly planar), and the analysis of stability rarely went beyond the basic definition. To obtain general results and a better understanding of these concepts, the present study addresses in a single general framework the questions of image likelihood and stability. We specifically eliminate irrelevant image transformations, such as rotation and scale, in the computation of image likelihood and stability (a factor that was not taken into explicit consideration before, unless the quantities involved were invariant to such transformations to begin with). Unlike the case with numerical simulations of specific objects, our analysis gives expressions that can be used to directly obtain the stability and likelihood profiles of general objects. Our analysis also gives a universal qualitative characterization of all objects, namely, uniquely identifying the most likely and stable views of an object. Our measure of stability may be considered a generalization to the Lie derivatives of the group of camera's rigid transformations, but unlike Lie derivatives it is free of distortions due to irrelevant image transformations and is not completely local.

The qualitative analysis identifies the most stable views of an object, which are the most suitable views to be used as the object's characteristic views in any of the approaches discussed above. Before anything else, however, our analysis gives a quantitative characterization of images, which by itself has a number of applications:

Witkin and Tenenbaum [31] argued for the importance of a Bayesian approach to object recognition (by emphasizing the importance of non-accidentalness of image properties). More formally, Freeman [12] suggested using a measure of view stability in the disambiguation of ambiguous situations. In his Bayesian scheme an interpretation which involves the more stable, or more generic, viewpoint is preferred. On a less formal level, in [24] the authors proposed a "Bayesian" approach to object recognition, where the probability is simply the invariant distance between an object and a model. Bayesian recognition and image understanding is one of the primary applications of our analysis: in order to select the most likely model from a library of models, each of which is a possible interpretation of an object in the scene (see Figs. 1, 2), the conditional distribution of images given models, derived here, is needed.

Another application of our analysis is the addition of quantitative characterization to a qualitative shape description, the aspect graph. The aspect graph, a coarse partition up to topological transformations of the different 2D views of a 3D object, was first defined in [18] and used for example in [13, 26, 29]. In this approach, the viewing sphere of an object is divided into separate bins, called aspects. The transition between aspects is characterized by topological events such as a change in occlusion, or the appearance and disappearance of features. Our analysis adds quantitative information on the "goodness" of each view in an aspect, and qualitative information on the "best" view in the aspect.

Our analysis can also be used to reduce the complexity of image-to-model correspondence. Various recognition methods of 3D objects require correspondence between a 2D image and a library of 3D models (e.g., alignment [15], classification by metric [2]). Image to model correspondence (or indexing) is computationally difficult, and may require exponential searches (e.g., [14]). One solution is to use 2D templates for the direct matching of 2D images, which may considerably reduce the complexity of search from $O(n^3)$ to $O(dn^2)$, where d is the number of templates (see [1] for a discussion of algorithms for finding all such matches). Our characterization of image likelihood and stability makes it possible to select the "best" templates, which can be matched to the largest amount of different views with the smallest amount of error. Moreover, we are able to identify sub configurations which are particularly stable and therefore should be relied on more heavily during the initial stage of correspondence.

2. Stability and Likelihood of Views: Definitions

In this section we define measures of image likelihood and stability. The likelihood measure depends on the conditional probability density function of images, which has a complex dependency on the prior conditional density of viewpoints. Each measure provides a somewhat different answer to a similar question: how representative is a particular two dimensional view of a three dimensional object?

2.1. The Viewing Sphere

The viewing sphere is an imaginary sphere around the centroid of the object, representing all the possible different viewpoints of the object. We assume weak perspective projection[3] and therefore the images of an object taken up to camera

---

[3] Weak perspective is an approximation to the pinhole perspective projection, which assumes that the field of view is not too large and therefore higher order terms in the size/distance ratio (the ratio between the size of the object and its distance to the camera) can be neglected. In this case, a world point whose coordinates in the camera coordinate system are (x, y, z) is projected to the image coordinates $s(x, y)$ for some scalar s, which is the same for all of the points (see [5] for a comparison between weak perspective and perspective projections).

translation and rotation around the camera's optical axis are equivalent up to a $2D$ similarity. Thus we treat all images that are equivalent up to a $2D$ similarity as the same. With these conventions, the viewing sphere describes all the possible different orientations of the camera with respect to the object. Therefore a view (or image) of the object corresponds to a point on the viewing sphere, which is completely defined by two angles. In the following, the viewing sphere is parameterized by a spherical coordinate system, where φ is the azimuth (longitude) and ϑ is the elevation (colatitude). Thus the range $\vartheta \in [0, \pi/2]$, $\varphi \in [0, 2\pi]$ parameterizes half the viewing sphere.

2.2. The Scope of the Analysis

The analysis in this paper describes general three dimensional objects, including opaque objects with occlusions. The only restriction is that salient points can be identified on the object. We consider separately parts of the object's viewing sphere according to the following characterization:

Opaque objects: one topological aspect.

Parts of objects: all the viewpoints from which some features belonging to the object part are visible (several topological aspects may be included).

Transparent objects: the whole viewing sphere of the object.

2.3. Stability and Likelihood of Views

Consider an object $\mathcal{O}$ and a point on the viewing sphere of $\mathcal{O}$ denoted by $V_{\vartheta,\varphi}$. Let $V'_{\vartheta,\varphi,\alpha,\beta}$ denote another view, corresponding to a rotation in spherical coordinates on the viewing sphere, where the point ϑ, φ is now the pole, $\alpha \in [0, \pi/2]$ the elevation, and $\beta \in [0, 2\pi]$ the azimuth (see Fig. 3). The geodesic distance from $V_{\vartheta,\varphi}$ to $V'_{\vartheta,\varphi,\alpha,\beta}$ on the viewing sphere is α. Let $d(\vartheta, \varphi, \alpha, \beta)$ denote the *image distance* between the images obtained from view V and view V'.

For each view $V = [\vartheta, \varphi]$ we measure the following:

Υ-variability: the maximal error (difference) d, when compared to other images obtained from views on the viewing sphere separated from it by less than Υ:

$$\max_{\alpha \leq \Upsilon, \beta} d(\vartheta, \varphi, \alpha, \beta) \tag{1}$$

Variability of an image measures the amount of change in the image when the camera moves Υ on the viewing sphere. Variability is inversely proportional to stability, and therefore the smaller the Υ-variability of a view is, the more stable it is.

ε-likelihood: the measure (on the viewing sphere) of the set $\{(\alpha, \beta)|$ such that $d(\vartheta, \varphi, \alpha, \beta) \leq \varepsilon\}$:

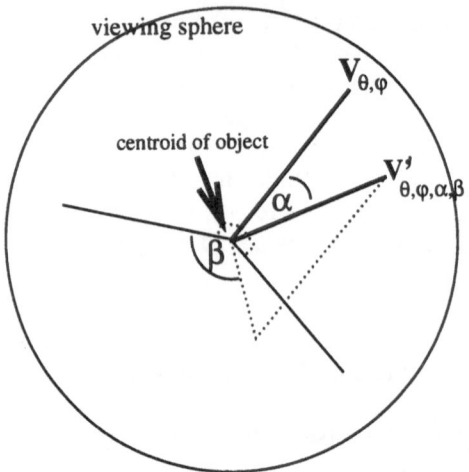

Figure 3. 2 views on the viewing sphere, $V_{\vartheta,\varphi}$ and $V'_{\vartheta,\varphi,\alpha,\beta}$

$$\int_{\{d(\vartheta,\varphi,\alpha,\beta)\leq\varepsilon\}} f(\alpha,\beta)\,d\alpha\,d\beta \tag{2}$$

where $f(\alpha,\beta)$ denotes the (prior) distribution of camera orientations relative to the object. (As an example, $f(\alpha,\beta) = 1$ when all the possible orientations of the camera are equally likely). Likelihood of an image is the probability of seeing another image that is almost (ε) the same.

The above metric characterization of views allows us to compare between them. We are also interested in two qualitative characterizations of views:

Most stable view: the view $V = [\vartheta,\varphi]$ which for all bounded movements of the viewing point from V the image changes the least:

$$\min_{\vartheta,\varphi} \max_{\alpha\leq\Upsilon,\beta} d(\vartheta,\varphi,\alpha,\beta) \tag{3}$$

Most likely view: the view $V = [\vartheta,\varphi]$ which has the largest number of views that are close to it as images:

$$\max_{\vartheta,\varphi} \int_{\{d(\vartheta,\varphi,\alpha,\beta)\leq\varepsilon\}} f(\alpha,\beta)\,d\alpha\,d\beta \tag{4}$$

3. Image Characterization and Comparison

To compute stability and likelihood of three dimensional objects, we need to be able to compare between their two dimensional images, obtained from different viewpoints on the viewing sphere. In this section we develop and review expressions for image comparison. We compare images by matching salient points, so that the only relevant parts of the object are the pre-images of these salient points, which is a set of n three dimensional fiducial points. Most importantly we do not wish to distinguish between images that differ by "irrelevant" image transformations, to be defined shortly.

3.1. Images of Objects with Fiducial Points

Let $\{\hat{\mathbf{p}}_i = (\hat{x}_i, \hat{y}_i, \hat{z}_i)\}_{i=1}^n$ denote the coordinates of the object features in the camera coordinate system in $\mathscr{R}^3$. A three dimensional representation of the object is the $3 \times n$ matrix $\hat{\mathbf{P}}$, whose i-th column is $\hat{\mathbf{p}}_i$, the vector describing the world coordinates of the i-th feature of the object.

An image of the object is obtained by a rigid transformation (of the object or the camera), followed by weak perspective (or scaled orthographic) projection from three dimensional space to the two dimensional image. An image of the object is therefore the set of n image points $\{\mathbf{p}_i = (x_i, y_i)\}_{i=1}^n$. An equivalent representation of the image is the $2 \times n$ matrix $\mathbf{P}$, whose i-th column is the image coordinates of the i-th feature of the object. The image matrix $\mathbf{P}$ is a sub-matrix of $\hat{\mathbf{P}}$ (up to a scaling factor), containing its first two rows.

The use of matrix $\mathbf{P}$ to represent an image of the object implies a correspondence between the image features and the object features, where different correspondences lead to permutations of the matrix' columns. Note that in this work we compare different images of the same object. For our analysis, therefore, the correspondence between images is given.

3.2. How to Compare Two Images

Given two images, or the two matrices $\mathbf{P}$ and $\mathbf{Q}$, the question of comparing them is equivalent to matrix comparison. We are using the "usual" metric, which is the Frobenius norm of the difference matrix, and which is the same as the Euclidean distance between points in the images:

$$\|\mathbf{P} - \mathbf{Q}\|_F^2 = \sum_{i,j} (\mathbf{P}[i,j] - \mathbf{Q}[i,j])^2 = tr[(\mathbf{P} - \mathbf{Q}) \cdot (\mathbf{P} - \mathbf{Q})^T] \tag{5}$$

(tr denotes the trace of a matrix). Henceforth we will omit the subscript F, and a matrix norm will be the Frobenius norm.

Before taking the norm of the difference between the images, we want to remove differences which are due to irrelevant effects, such as the size of the image (which is arbitrary under scaled orthography) or the exact location of the object (e.g., due to an arbitrary translation and rotation of the object in the image). In particular, we may want to consider as equivalent all images obtained from each other by one of the following two groups of two dimensional transformations: the **similarity** group, which includes 2D rotations, translations, and scale, or the **affine** group, which includes 2D linear transformations and translations.

The equivalence under similarity transformations is necessary, since under weak perspective projection, images that differ by image scale, rotation or translation can be obtained from the same object, and should therefore be considered the *same image*. Since the group of affine transformations includes the similarity group as a sub-group, the equivalence under affine transformations is more general: it makes images that differ by a 2D rotation and scale appear the same as needed, but it also makes images of different objects appear the same. It therefore leads to false identifications of different images as the same image. On the other hand, for planar objects, the affine alignment makes *all* the images of the object (from all viewpoints) appear the same, which is advantageous when planar objects are expected. Both measures are therefore useful, and we consider both possibilities here.

It can be readily shown that the optimal translation when measuring distance by sum of square distances, under both the similarity and affine equivalence, puts the centroid of the object in the origin of the image. We therefore assume w.l.o.g. that the images are centered on the centroid of the object, so that the first moments of the objects are 0. In [30] we define image distance measures, which satisfy all the properties of a metric, and which compare the images $\mathbf{P}$ and $\mathbf{Q}$ while taking into account the desired image equivalence discussed above. We get the following expressions for the similarity metric $D_{sim}(\mathbf{P}, \mathbf{Q})$ and the affine metric $D_{aff}(\mathbf{P}, \mathbf{Q})$:

$$D_{sim}^2(\mathbf{P}, \mathbf{Q}) = 1 - \frac{\|\mathbf{QP}^T\|^2 + 2\,det(\mathbf{QP}^T)}{\|\mathbf{P}\|^2 \|\mathbf{Q}\|^2} \tag{6}$$

$$D_{aff}^2(\mathbf{P}, \mathbf{Q}) = 2 - tr(\mathbf{P}^+\mathbf{P} \cdot \mathbf{Q}^+\mathbf{Q})$$

($A^+ = (A^T A)^{-1} A^T$ denotes the pseudo-inverse of a matrix A).

The above metrics can be obtained by taking the difference between the two images after a two stage compensation process, where the images are first normalized by a scaling (similarity) or by moment normalization (affine), and then one image is optimally aligned with the other with a rotation (similarity) or a linear transformation (affine). However, under orthographic projection the scale of the image is known, and we want to avoid normalization and instead compare the model to the image as is. We therefore take the difference between the given image and the model, where the model is aligned with an affine transformation to the image, and without any manipulation permitted to be applied to the image. This asymmetric treatment of the given image and the stored image (model) may be appropriate in the context of a recognition task, where we need to select the model that best fits a given image, and therefore we should manipulate the model to best fit the image as is. If we denote the model $\mathbf{P}$ and the new image $\mathbf{Q}$, we define the following orthographic distance measure:

$$D_{ortho}^2(\mathbf{P}, \mathbf{Q}) = \min_A \|A\mathbf{P} - \mathbf{Q}\|^2 = tr[\mathbf{Q}^T\mathbf{Q}(I - \mathbf{P}^+\mathbf{P})] \tag{7}$$

(I denotes the $n \times n$ unity matrix). Note that D_{ortho} is not a metric as it is not even symmetric.

So far we described how to compare images when only the image coordinates are given. This is the typical case in image understanding. However, for the analysis of stability and likelihood of views of known objects, which is the focus of this paper, we will now show how to compute the above distance from the *world* three dimensional coordinates of the points, matrices $\hat{\mathbf{P}}$ and $\hat{\mathbf{Q}}$. For simplicity of notation, denote by $(Y)_{2x2}$ the upper-left 2×2 sub-matrix of a 3×3 matrix Y. Modifying Eq. (6) and Eq. (7) we get:

$$D^2_{sim}(\mathbf{P},\mathbf{Q}) = 1 - \frac{\|(\hat{\mathbf{Q}}\hat{\mathbf{P}}^T)_{2x2}\|^2 + 2\,det((\hat{\mathbf{Q}}\hat{\mathbf{P}}^T)_{2x2})}{tr((\hat{\mathbf{P}}\hat{\mathbf{P}}^T)_{2x2})\,tr((\hat{\mathbf{Q}}\hat{\mathbf{Q}}^T)_{2x2})}$$

$$D^2_{aff}(\mathbf{P},\mathbf{Q}) = 2 - tr[((\hat{\mathbf{Q}}\hat{\mathbf{P}}^T)_{2x2})((\hat{\mathbf{P}}\hat{\mathbf{P}}^T)_{2x2})^{-1}(\hat{\mathbf{P}}\hat{\mathbf{Q}}^T)_{2x2}((\hat{\mathbf{Q}}\hat{\mathbf{Q}}^T)_{2x2})^{-1}] \quad (8)$$

$$D^2_{ortho}(\mathbf{P},\mathbf{Q}) = tr[(\hat{\mathbf{Q}}\hat{\mathbf{Q}}^T)_{2x2} - (\hat{\mathbf{Q}}\hat{\mathbf{P}}^T)_{2x2}((\hat{\mathbf{P}}\hat{\mathbf{P}}^T)_{2x2})^{-1}(\hat{\mathbf{P}}\hat{\mathbf{Q}}^T)_{2x2}]$$

Here we used the fact that $\hat{\mathbf{P}}$ and $\hat{\mathbf{Q}}$ contain the first two rows of $\mathbf{P}$ and $\mathbf{Q}$ respectively up to a scale factor. We can ignore the differences in scale when computing D_{sim} and D_{aff} since these metrics implicitly include normalization of the images.

3.3. The "Flattest" View

For each view of the object V, let $\hat{\mathbf{P}}_V$ denote the $3D$ coordinates of the object in a viewer-centered coordinate system, where the image is the $X-Y$ plane, and the viewing direction is the Z axis. Let S_V denote the 3×3 symmetric autocorrelation scatter matrix of the object: $S_V = \hat{\mathbf{P}}_V \cdot \hat{\mathbf{P}}_V^T$.

Definition 1. The **flattest view** is a view V_f whose scatter matrix S_{V_f} is diagonal, and where the eigenvalues (the diagonal elements) are ordered in decreasing order. Let S_0 denote the diagonal scatter matrix at the flattest view:

$$S_0 = S_{V_f} = \begin{bmatrix} a & 0 & 0 \\ 0 & b & 0 \\ 0 & 0 & c \end{bmatrix}$$

where $a \geq b \geq c > 0$ (a, b, c are the 3 principal second moments of the object).

Recall that a symmetric matrix can always be diagonalized by a similarity transformation with an orthogonal matrix. Such a diagonalization of the scatter matrix S_V is equivalent to a rotation of the coordinate system defining $\hat{\mathbf{P}}_V$, namely, a change of viewpoint. Thus for any object there always exists a view which is the flattest view of the object. This view can be computed from S_V of any V, by computing the orthogonal matrix which diagonalizes S_V. It is unique if $b > c$. Recall that a view is specified by 2 angles ϑ, φ, parameterizing the viewing sphere of the object. Henceforth we will assume w.l.o.g. that the viewing sphere is initially parameterized so that $V_{\vartheta=0,\varphi=0} = V_f$.

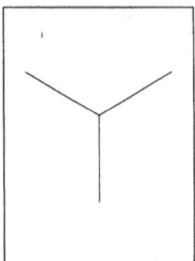

Figure 4. The flattest view of a straight corner

As an example, consider a three dimensional straight corner, an object composed of the points: $\{(0,0,0),(1,0,0),(0,1,0),(0,0,1)\}$. After centering this object, its three principal second moments are $a = 1, b = 1, c = 0.25$ (they are not all 1). The flattest view of this object is shown in Fig. 4.

4. Metric Characterization of Views

In this section we develop the expressions for likelihood and stability of views, as a function defined over the viewing sphere. We obtain closed-form expressions that make it possible to instantly obtain and plot these functions for every given object. As an example, we compute the stability and likelihood of the images of a pyramid.

4.1. The Distance between Two Views

Consider an object containing n fiducial points in three dimensional space, $\mathcal{O} = \hat{\mathbf{p}}_1, \hat{\mathbf{p}}_2, \ldots, \hat{\mathbf{p}}_n$. Let $\hat{\mathbf{P}}$ denote the $3 \times n$ matrix whose i-th column is $\hat{\mathbf{p}}_i$.

As defined in Section 2.3, let $V_{\vartheta,\varphi}$ denote a point on the viewing sphere of object $\mathcal{O}$, and let $V'_{\vartheta,\varphi,\alpha,\beta}$ denote another view of $\mathcal{O}$. Let $d(\vartheta, \varphi, \alpha, \beta)$ denote the image distance, which is one of the three distance measures defined in Section 3.2, between the appearance of object $\mathcal{O}$ from view V and its appearance from view V'. For all three image distance measures we can show that:

Result 1. $d(\vartheta, \varphi, \alpha, \beta)$ depends only on the 3 principal second moments (a, b, c) of the object, regardless of the number of features in $\mathcal{O}$ or their distribution in space. We therefore denote the distance by $d(\vartheta, \varphi, \alpha, \beta; a, b, c)$.

This result is remarkable. It shows that the 3 principal second moments of an object completely characterize the stability and likelihood of each of its views, regardless of the particular shape of the object. Note that this is a result, and not an assumption, of our analysis. The result demonstrated in Fig. 5a.

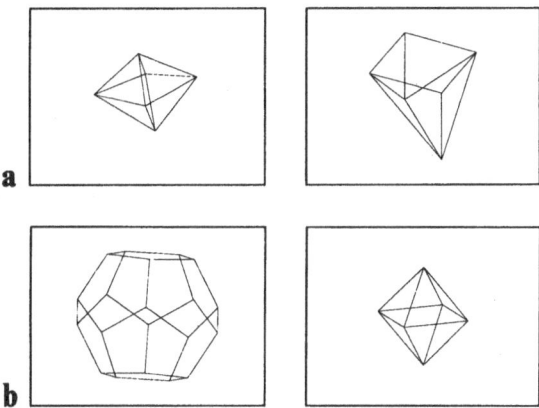

Figure 5. a Transparent octahedron and dodecahedron, both having three identical principal second moments. **b** Two object composed of six points, and having the same principal second moments: 8, 6, 4. The distance between any two images of the objects in **a** or **b**, taken from the same viewpoints, is the same

We computed $d(\vartheta, \varphi, \alpha, \beta; a, b, c)$ for the two metrics defined in Eq. (6)[4]. For the similarity metric we get:

$$D_{sim}^2(\vartheta, \varphi, \alpha, \beta; a, b, c) = \frac{(1 - \cos(\alpha))(abs_1 + acs_2 + bcs_3)}{u(at_1 + bt_2 + ct_3)} \qquad (9)$$

where

$$s_1 = 1 - 2\cos^2(\vartheta)\cos(\alpha) + 2\cos(\vartheta)\sin(\alpha)\sin(\vartheta)\cos(\beta) + \cos(\alpha)$$

$$s_2 = 1 - 2\cos(\alpha)\cos^2(\varphi)\sin^2(\vartheta) + \cos(\alpha) - 2\cos(\beta)\cos^2(\varphi)\sin(\alpha)\sin(\vartheta)\cos(\vartheta)$$
$$+ 2\sin(\beta)\sin(\alpha)\sin(\varphi)\cos(\varphi)\sin(\vartheta)$$

$$s_3 = 1 + 2\cos(\alpha)\cos^2(\varphi) + 2\cos(\alpha)\sin^2(\varphi)\cos^2(\vartheta)$$
$$- 2\sin(\beta)\sin(\alpha)\sin(\varphi)\cos(\varphi)\sin(\vartheta)$$
$$- 2\cos(\beta)\sin^2(\varphi)\sin(\alpha)\sin(\vartheta)\cos(\vartheta) - \cos(\alpha)$$

$$u = a(1 - \sin^2(\varphi)\sin^2(\vartheta)) + b(1 - \cos^2(\varphi)\sin^2(\vartheta)) + c\sin^2(\vartheta)$$

$$t_1 = -2\cos(\alpha)\cos(\beta)\sin(\alpha)\cos(\vartheta)\sin(\vartheta)\sin^2(\varphi)$$
$$- 2\cos(\varphi)\sin(\varphi)\cos(\vartheta)\sin^2(\alpha)\cos(\beta)\sin(\beta) - \cos^2(\vartheta)\sin^2(\alpha)\cos^2(\beta)\sin^2(\varphi)$$
$$+ 1 - 2\cos(\varphi)\cos(\alpha)\sin(\beta)\sin(\alpha)\sin(\vartheta)\sin(\varphi) - \sin^2(\vartheta)\cos^2(\alpha)\sin^2(\varphi)$$
$$- \cos^2(\varphi)\sin^2(\alpha)\sin^2(\beta)$$

---

[4] In deriving the final form, we used the symbolic package Maple.

$t_2 = 2\cos(\varphi)\sin(\varphi)\cos(\vartheta)\sin^2(\alpha)\cos(\beta)\sin(\beta) + \sin^2(\varphi)\sin^2(\alpha)\cos^2(\beta)$

$\quad + \cos^2(\varphi)\sin^2(\alpha) + \cos^2(\vartheta)\cos^2(\alpha)\cos^2(\varphi)$

$\quad + 2\cos(\varphi)\cos(\alpha)\sin(\beta)\sin(\alpha)\sin(\vartheta)\sin(\varphi) + \cos^2(\alpha)\sin^2(\varphi)$

$\quad - 2\cos(\alpha)\cos(\beta)\sin(\alpha)\cos(\vartheta)\sin(\vartheta)\cos^2(\varphi) - \cos^2(\vartheta)\sin^2(\alpha)\cos^2(\beta)\cos^2(\varphi)$

$t_3 = 1 + 2\cos(\vartheta)\sin(\alpha)\sin(\vartheta)\cos(\alpha)\cos(\beta) - \cos^2(\vartheta)\cos^2(\alpha)$

$\quad - \sin^2(\vartheta)\sin^2(\alpha)\cos^2(\beta)$

For the affine metric we get:

$$D_{aff}^2(\vartheta, \varphi, \alpha, \beta; a, b, c) = abc\,\frac{\sin^2(\alpha)(as_1 + bs_2 + cs_3)}{u(abt_1 + act_2 + bct_3)} \tag{10}$$

where

$s_1 = (\cos(\beta)\cos(\varphi) - \sin(\beta)\cos(\vartheta)\sin(\varphi))^2$

$s_2 = (\cos(\beta)\sin(\varphi) + \sin(\beta)\cos(\vartheta)\cos(\varphi))^2$

$s_3 = \sin^2(\beta)\sin^2(\vartheta)$

$u = ab\cos^2(\vartheta) + ac\cos^2(\varphi)\sin^2(\vartheta) + bc\sin^2(\varphi)\sin^2(\vartheta)$

$t_1 = (\sin(\alpha)\cos(\beta)\sin(\vartheta) - \cos(\alpha)\cos(\vartheta))^2$

$t_2 = (\cos(\alpha)\sin(\vartheta)\cos(\varphi) - \sin(\alpha)\sin(\beta)\sin(\varphi))^2 + \sin^2(\alpha)\cos^2(\beta)\cos^2(\vartheta)\cos^2(\varphi)$

$\quad - 2\cos(\beta)\sin(\beta)\cos(\vartheta)\cos(\varphi)\sin^2(\alpha)\sin(\varphi)$

$\quad + 2\cos(\alpha)\cos(\beta)\cos^2(\varphi)\sin(\alpha)\sin(\vartheta)\cos(\vartheta)$

$t_3 = 2\cos(\alpha)\sin(\alpha)\sin(\varphi)\cos(\varphi)\sin(\beta)\sin(\vartheta)$

$\quad + 2\cos(\alpha)\cos(\beta)\sin^2(\varphi)\sin(\alpha)\sin(\vartheta)\cos(\vartheta)$

$\quad + \sin^2(\alpha)(\cos(\beta)\cos(\vartheta)\sin(\varphi) + \sin(\beta)\cos(\varphi))^2 + \cos^2(\alpha)\sin^2(\vartheta)\sin^2(\varphi)$

The expression for the orthographic distance is remarkably simple:

$$D_{ortho}^2(\vartheta, \varphi, \alpha, \beta; a, b, c) = \frac{abc\sin^2(\alpha)}{ac\sin^2(\vartheta)\cos^2(\varphi) + bc\sin^2(\vartheta)\sin^2(\varphi) + ab\cos^2(\vartheta)} \tag{11}$$

4.2. Stability and Likelihood at Each View

Using the definitions given in Section 2.3, we can now compute the stability and likelihood of each view, corresponding to a point on the viewing sphere $[\vartheta, \varphi]$. More specifically, let $v(\vartheta, \varphi)$ denote the variability function, and $l(\vartheta, \varphi)$ denote the likelihood function. (1) and (2) become:

$$v_m(\vartheta, \varphi; a, b, c) = \max_{\alpha \leq \Upsilon, \beta} D_m^2(\vartheta, \varphi, \alpha, \beta; a, b, c) \tag{12}$$

$$l_m(\vartheta, \varphi; a, b, c) = \int_{\{D_m^2(\vartheta, \varphi, \alpha, \beta; a, b, c) \leq \varepsilon\}} f(\alpha, \beta) \, d\alpha \, d\beta \tag{13}$$

where $f(\alpha, \beta)$ denotes the (prior) distribution of camera orientations relative to the object. The subscript m denotes the type of image metric used: similarity, affine, or orthographic. We obtain 6 functions: $v_{sim}(\vartheta, \varphi; a, b, c)$, $v_{aff}(\vartheta, \varphi; a, b, c)$, $v_{ortho}(\vartheta, \varphi; a, b, c)$, $l_{sim}(\vartheta, \varphi; a, b, c)$, $l_{aff}(\vartheta, \varphi; a, b, c)$, $l_{ortho}(\vartheta, \varphi; a, b, c)$,

Using (11) the expressions for the orthographic distance measure can be greatly simplified, to obtain:

$$v_{ortho}(\vartheta, \varphi; a, b, c) = \frac{abc \sin(\Upsilon)^2}{ac \sin^2(\vartheta) \cos^2(\varphi) + bc \sin^2(\vartheta) \sin^2(\varphi) + ab \cos^2(\vartheta)} \tag{14}$$

$$l_{ortho}(\vartheta, \varphi; a, b, c)$$
$$= 1 - \frac{\sqrt{abc - \varepsilon ac \sin^2(\vartheta) \cos^2(\varphi) - \varepsilon bc \sin^2(\vartheta) \sin^2(\varphi) - \varepsilon ab \cos^2(\vartheta)}}{\sqrt{abc}} \tag{15}$$

In the derivation of (15) we used the uniform prior distribution over the viewing sphere.

4.3. Application: Characterization of an Aspect of a Pyramid

The equations developed above enable us to readily compute the likelihood and stability functions for any object composed of feature points. As an example, we will now develop these functions for a square (non-transparent) pyramid, whose nodes are at $\{(0, 0, 2), (1, 0, 0) (0, 1, 0), (-1, 0, 0), (0, -1, 0)\}$.

First, as we are dealing with a non-transparent object, we should carry out the analysis separately for each aspect of the object. By aspect we specifically mean all the views of the object from which the same features are visible. We will analyze one aspect where 4 feature points, 3 of the basis nodes $\{(1, 0, 0), (0, 1, 0), (-1, 0, 0)\}$ and the top of the pyramid $(0, 0, 2)$, are visible. One such image, which is the flattest view of the pyramid in this aspect, is given in Fig. 7. Without loss of generality we set the origin of the coordinate system parametrizing the viewing sphere, $(\vartheta = 0, \varphi = 0)$, to be at this flattest view.

We proceed as follows:

1. We center the visible feature points, $\{(0, 0, 2), (1, 0, 0), (0, 1, 0), (-1, 0, 0)\}$ at the origin, and compute their 3 principal second moments: $a = 3.1, b = 2, c = 0.64$.

2. We substitute these 3 second moments into (9)–(11) to obtain $D_{sim}(\vartheta, \varphi, \alpha, \beta;$ pyramid), $D_{aff}(\vartheta, \varphi, \alpha, \beta;$ pyramid), and $D_{ortho}(\vartheta, \varphi, \alpha, \beta;$ pyramid) respectively.

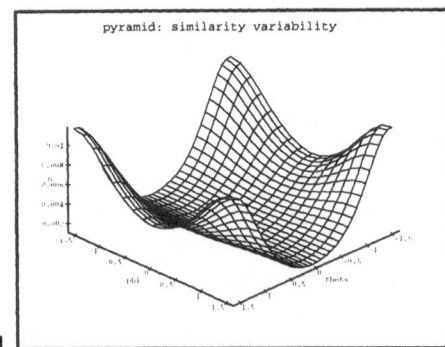
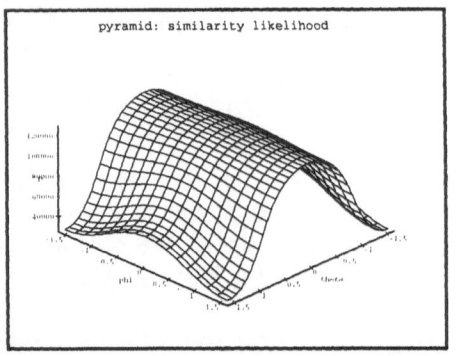

Figure 6. a variability function; **b** likelihood function for the similarity metric. Note that although the functions are plotted over the range of ϑ, φ which spans half the viewing sphere, they are only defined over ϑ, φ which are at the same aspect as $\vartheta = 0$, $\varphi = 0$

3. We substitute D_{sim}, D_{aff}, and D_{ortho} into (12)–(15) to obtain 6 functions:

- $v_{sim}(\vartheta, \varphi; pyramid)$—the variability at view $[\vartheta, \varphi]$ of the pyramid under the similarity metric (Fig. 6—left).
- $l_{sim}(\vartheta, \varphi; pyramid)$—the likelihood at view $[\vartheta, \varphi]$ of the pyramid under the similarity metric (Fig. 6 right).
- $v_{aff}(\vartheta, \varphi; pyramid)$—the variability at view $[\vartheta, \varphi]$ of the pyramid under the affine metric.
- $l_{aff}(\vartheta, \varphi; pyramid)$—the likelihood at view $[\vartheta, \varphi]$ of the pyramid under the affine metric.
- $v_{ortho}(\vartheta, \varphi; pyramid)$—the variability at view $[\vartheta, \varphi]$ of the pyramid under the orthographic metric.
- $l_{ortho}(\vartheta, \varphi; pyramid)$—the likelihood at view $[\vartheta, \varphi]$ of the pyramid under the orthographic metric.

We used $\varepsilon = 0.2$ and $Y = 0.2$. In obtaining the likelihood functions, we assumed that all views of the pyramid are equally likely (i.e., a uniform prior viewpoint distribution).

The variability function, shown in Fig. 6, is rather insensitive to the particular image metric used. In particular, this function is almost indistinguishable under the three image distances. Therefore (14) and (15) can be used to approximate the variability and likelihood functions, respectively, with the benefit of simplicity at the expense of some inaccuracy.

5. Qualitative Characterization of Views

Using the variability and likelihood functions at each view, as computed above, we can now compute a qualitative characterization of aspects on the viewing sphere.

5.1. The Most Stable and Likely View

For each aspect, we will compute:

- The most likely view—the view which obtains the maximum of the likelihood function over all the views in the aspect.
- the most stable view—the view which obtains the minimum of the variability function over all the views in the aspect.

We computed the most stable and most likely views for various objects, characterized by different principal second moments a, b, c, and for various likelihood and stability thresholds ε and $\varUpsilon$. The simulations lead us to conjecture the following result (which can be proved for the orthographic distance):

Result 2. The flattest view $V_f = V_{0,0}$ is both the $\varUpsilon$-variable view and the ε-likely view (in the case of a uniform prior distribution on the viewing sphere) for all $\varUpsilon$ and ε, and for every object.

This result may seem intuitive for the affine and orthographic distance measures (although it is less intuitive for the similarity one), as a perfectly flat (2D) object is affinely equivalent from every viewpoint.

5.2. Examples

As an example, we computed V_f for three familiar objects: a cube, a pyramid, and a straight corner. For the cube we considered an aspect where 7 vertices are visible. For the pyramid we considered an aspect where 4 vertices are visible (as in the example discussed above in Section 4.3). Figure 7 shows the V_f of each of these objects. Figure 4 shows the V_f of a straight corner.

Note that the result for the pyramid can be seen directly in Fig. 6, where the minimum of $l_{sim}(\vartheta, \varphi)$ and the maximum of $v_{sim}(\vartheta, \varphi)$, are always obtained at $\vartheta = 0$, $\varphi = 0$, which is the flattest view by our definition of the coordinate system.

5.3. Stronger Results for the Orthographic Distance

We saw that the expression for D_{ortho} is much simpler than D_{aff} and D_{sim}. We also saw in the pyramid example that the resulting likelihood and variability functions are rather similar when the affine and orthographic image distances are used. Thus the orthographic case teaches us about the more general affine case, and we therefore derive additional results for the orthographic distance measure, using (14) and (15):

- The ε-likelihood of V_f is: $1 - \sqrt{\dfrac{c - \varepsilon}{c}}$.
- The $\varUpsilon$-variability of V_f is: $c \sin^2(\varUpsilon)$.

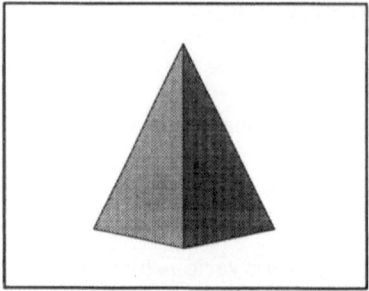 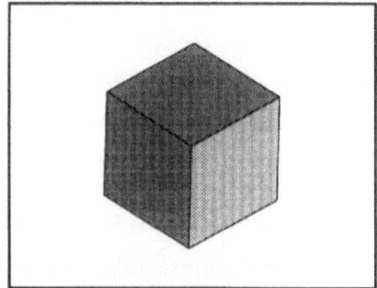

Figure 7. The flattest views of two opaque objects: a square pyramid and a cube

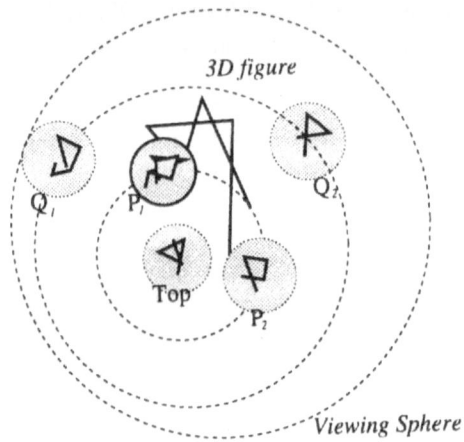

Figure 8. Different views of a 3D object on the viewing sphere. The views marked by P_i are at the same elevation from *Top*, and thus at the same orthographic distance from *Top*; likewise, the Q's are at the same orthographic distance from *Top*, which is larger than the orthographic distance of the P's

From (14), (15), and the above expressions we get the following extension of result 2:

Result-ext 2. Properties of the flattest view:

1. V_f is the only view which is locally the most likely and the most stable when compared to views in its immediate neighborhood. This follows from the fact that the likelihood and variability surfaces have a single local minimum/maximum.
2. The difference between $V_{0,0}$ and $V_{0,0,\alpha,\beta}$ is $c \sin^2(\alpha)$.
3. For any α, β, the distance from $V_{0,0}$ to $V_{0,0,\alpha,\beta}$ is less than the distance from $V_{\vartheta,\varphi}$ to $V_{\vartheta,\varphi,\alpha,\beta}$.

In addition, we get the following somewhat surprising result for the orthographic metric:

Result 3. For all ϑ, φ, the distance between $V_{\vartheta,\varphi}$ and $V_{\vartheta,\varphi,\alpha,\beta}$ depends only on the geodesic distance α and does not depend on the azimuth β, although for different azimuth the views $V_{\vartheta,\varphi,\alpha,\beta}$ are not affine equivalent.

In other words, if we fix a view V as the pole on the viewing sphere, all the viewpoints that are on the same latitude on the viewing sphere induce images which are at the same distance from the image of V (see Fig. 8).

Acknowledgements

This research was sponsored by the U.S. Office of Naval Research under Grant N00014-93-1-1202, R&T Project Code 4424341-01. DW and NT were also supported by the Israeli Science Foundation grant 202/92-2. Part of this research was done while DW and MW visited the Newton Institute of Cambridge University, in July 1993.

References

[1] Arkin, E. M., Kedem, K., Mitchell, J. S. B., Sprinzak, J., Werman, M.: Matching points into pairwise disjoint noise regions: Combinatorial bounds and algorithms. ORSA J. Comput. (1992).
[2] Basri, R., Weinshall, D.: Distance metric between 3D models and 2D images for recognition and classification. IEEE Trans. Pattern Anal. Machine Intell., 1995 (to appear).
[3] Ben-Arie, J.: The probabilistic peaking effect of viewed angles and distances with application to 3-D object recognition. T-PAMI *12*, pp. 760–774 (1990).
[4] Binford, T. O., Levitt, T. S.: Quasi-invariants: theory and exploitation. Image Understanding Workshop, pp. 819–829, 1993.
[5] Boufama, B., Weinshall, D., Werman, M.: Shape from motion algorithms: a comparative analysis of scaled orthography and perspective. In: ECCV, pp. 199–204, Stockholm, Sweden, 1994.
[6] Bülthoff, H. H., Edelman, S.: Psychophysical support for a 2D interpolation theory of object recognition. Proc. Natl. Acad. Sci. USA *89*, 60–64 (1992).
[7] Burns, J. B., Weiss, R., Riseman, E.: View variation of point-set and line segment features. IEEE Trans. Pattern Anal. Machine Intell. *15*, 51–68 (1993).
[8] Chakravarty, I., Freeman, H.: Characteristic views as a basis for three-dimensional object recognition. In: Proc. SPIE Conf. Robot Vision *336*, 37–45 (1982).
[9] Dickinson, S. J., Pentland, A. P., Rosenfeld, A.: 3-D shape recovery using distributed aspect matching. IEEE Trans. Pattern Anal. Machine Intell. *14*, 174–198 (1992).
[10] Edelman, S., Weinshall, D.: A self-organizing multiple-view representation of 3D objects. Biol. Cybernetics *64*, 209–219 (1991).
[11] Forsyth, D., Mundy, J. L., Zisserman, A., Coelho, C., Heller, A., Rothwell, C.: Invariant descriptors for 3-D object recognition and pose. IEEE Trans. Pattern Anal. Machine Intell. *13*, 971–991 (1991).
[12] Freeman, W. T.: Exploiting the generic view assumption to estimate scene parameters. In: Proceedings of the 4th International Conference on Computer Vision, pp. 347–356. Berlin, Germany. Washington, DC: IEEE 1993.
[13] Gigus, Z., Malik, J.: Computing the aspect graph for line drawings of polyhedral objects. T-PAMI *12*, 113–122 (1990).
[14] Grimson, W. E. L.: The effect of indexing on the complexity of object recognition. In: Proceedings of the 3rd International Conference on Computer Vision, pp. 644–651, Osaka, Japan. Washington, DC: IEEE 1990.
[15] Huttenlocher, D. P., Ullman, S.: Object recognition using alignment. In: Proceedings of the 1st International Conference on Computer Vision, pp. 102–111, London, England, June 1987. Washington, DC: IEEE 1987.
[16] Kanatani, K.: Group theoretical methods in image understanding. Berlin: Springer 1990.
[17] Keren, D.: Using symbolic computation to find algebraic invariants. T-PAMI *16*, 1143–1149 (1994).

[18] Koenderink, J. J., van Doorn, A. J.: The internal representation of solid shape with respect to vision. Biol. Cybernetics *32*, 211–217 (1979).
[19] Lamdan, Y., Wolfson, H.: Geometric hashing: a general and efficient recognition scheme. In: Proceedings of the 2nd International Conference on Computer Vision, pp. 238–251, Tarpon Springs, FL. Washington, DC: IEEE 1988.
[20] Moses, Y., Ullman, S.: Limitations of non model-based schemes. A.I. Memo No. 1301, Artificial Intelligence Laboratory, Massachusetts Institute of Technology, 1991.
[21] Newell, F., Findlay, J. M.: The effect of familiarity on the time to recognise depth-rotated objects. Perception [Suppl.] *22*, 226 (1993).
[22] Palmer, S. E., Rosch, E., Chase, P.: Canonical perspective and the perception of objects. In: Attention and Performance IX (Long, J., Baddeley, A., eds.), pp. 135–151. Hillsdale: Erlbaum 1981.
[23] Poggio, T., Edelman, S.: A network that learns to recognize three-dimensional objects. Nature *343*, 263–266 (1990).
[24] Rigoutsos, I., Hummel, R.: Distributed Bayesian object. In: Proceedings IEEE Conf. on Computer Vision and Pattern Recognition, pp. 180–186, 1993.
[25] Rock, I., DiVita, J.: A case of viewer-centered object perception. Cogn. Psychol. *19*, 280–293 (1987).
[26] Stark, L., Eggert, D., Bowyer, K.: Aspect graphs and nonlinear optimization in 3-D object recognition. In: Proceedings of the 2nd International Conference on Computer Vision, pp. 501–507, Tarpon Springs, FL. Washington, DC: IEEE 1988.
[27] Ullman, S., Basri, R.: Recognition by linear combinations of models. IEEE Trans. Pattern Anal. Machine Intell. *13*, 992–1006 (1991).
[28] Weiss, I.: Projective invariants of shapes. In: Proceedings IEEE Conf. on Computer Vision and Pattern Recognition, pp. 291–297, June 1988.
[29] Werman, M., Baugher, E. S., Gualtieri, A.: The visual potential: One convex polygon. Comput. Vision Graphics Image Proc. *45*, 96–130 (1989).
[30] Werman, M., Weinshall, D.: Similarity and affine distance between point sets. IEEE Trans. Pattern Anal. Machine Intell. (1995).
[31] Witkin, A. P., Tenenbaum, J. M.: On the role of structure in vision. In: Human and machine vision (Beck, J., Hope, B., Rosenfeld, A., eds.), pp. 481–544. New York: Academic Press 1983.

D. Weinshall, M. Werman, N. Tishby
Institute of Computer Science
The Hebrew University of Jerusalem
91904 Jerusalem
Israel
e-mail: daphna@cs.huji.ac.il

SpringerComputerScience

Hans Hagen, Gerald Farin,
Hartmut Noltemeier (eds.)

Geometric Modelling

Dagstuhl 1993

in cooperation with Rudolf Albrecht
1995. 188 figures. VII, 361 pages.
Soft cover DM 180,–, öS 1260,–
Reduced price for subscribers to "Computing":
Soft cover DM 162,–, öS 1134,–
ISBN 3-211-82666-1
Computing/Supplement 10

Experts from university and industry are presenting new technologies for solving industrial problems and giving many important and practicable impulses for new research. Topics explored include NURBS, product engineering, object oriented modelling, solid modelling, surface interrogation, feature modelling, variational design, scattered data algorithms, geometry processing, blending methods, smoothing and fairing algorithms, spline conversion.
This collection of 24 articles gives a state-of-the-art survey of the relevant problems and issues in geometric modelling.

 SpringerWienNewYork

P.O.Box 89, A-1201 Wien • New York, NY 10010, 175 Fifth Avenue
Heidelberger Platz 3, D-14197 Berlin • Tokyo 113, 3-13, Hongo 3-chome, Bunkyo-ku

SpringerMathematics

Rudolf Albrecht, Götz Alefeld,
Hans J. Stetter (eds.)

Validation Numerics

Theory and Applications

1993. 23 figures. IX, 291 pages.
Soft cover DM 160,–, öS 1120,–
Reduced price for subscribers to "Computing":
Soft cover DM 144,–, öS 1008,–
ISBN 3-211-82451-0
Computing/Supplement 9

The articles in this book give a comprehensive overview on
the whole field of validated numerics. The problems covered
include simultaneous systems of linear and nonlinear equations, differential and integral equations and certain applications from technical sciences. Furthermore some papers
which improve the tools are included. The book is a must for
scientists working in numerical analysis, computer science
and in technical fields.

 SpringerWienNewYork

P.O.Box 89, A-1201 Wien • New York, NY 10010, 175 Fifth Avenue
Heidelberger Platz 3, D-14197 Berlin • Tokyo 113, 3-13, Hongo 3-chome, Bunkyo-ku

Springer-Verlag and the Environment